GRASP

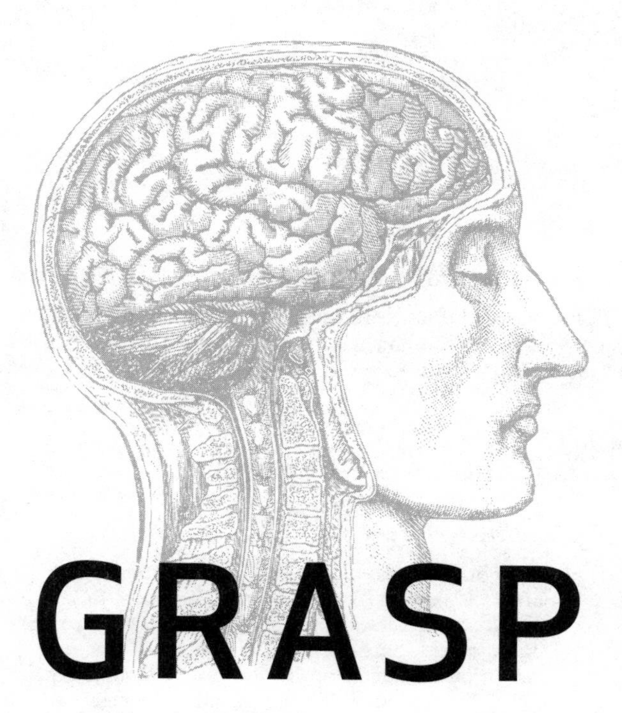

GRASP

The Science Transforming How We Learn

Sanjay Sarma

with Luke Yoquinto

DOUBLEDAY / NEW YORK

 Published in the United States by Doubleday, a division of Penguin Random House LLC, New York, and distributed in Canada by Penguin Random House Canada Limited, Toronto.

www.doubleday.com

Book design by Michael Collica
Jacket design and photography by Pete Garceau

Library of Congress Cataloging-in-Publication Data
Names: Sarma, Sanjay E., [date] author. | Yoquinto, Luke, author.
Title: Grasp : the science transforming how we learn /
Sanjay Sarma and Luke Yoquinto.
Description: First edition. | New York : Doubleday,
[2020] | Includes bibliographical references.
Identifiers: LCCN 2019054620 | ISBN 9780385541824 (hardcover) |
ISBN 9780385541831 (ebook) | ISBN 9780385546683 (open market)
Subjects: LCSH: Learning, Psychology of. | Learning—Research.
Classification: LCC LB1060 .S26 2020 | DDC 370.15/23—dc23
LC record available at https://lccn.loc.gov/2019054620

MANUFACTURED IN THE UNITED STATES OF AMERICA

1 3 5 7 9 10 8 6 4 2

First Edition

SS: To Gitanjali and Tara

LY: To Kelsey and Hope

In memory of Woodie Flowers

Contents

Introduction: The Adventure Begins

Learning is an adventure—and as anyone who's survived an adventure will tell you, that's not always a good thing. It can be a long, rewarding process that changes the way you experience the world and reveals capabilities you never knew you had. But in the retelling adventures are extolled only by the victors. For every adventurer who's slain an educational dragon, there may be dozens lying in caves and ditches nearby, smoldering in their armor, ruing the day they made their attempt.

My closest scrape with outrageous educational misfortune came at the Indian Institute of Technology, Kanpur. The moment stands out in my memory with uncanny clarity, as though it had been filmed at double the standard frame rate, as in Peter Jackson's somehow-too-vivid *Hobbit* movies. It happened my final year at university.

Actually, to be technical, it was the summer after my final year. Although I had entered university firing on all cylinders, at some point my straight As gave way to curvier letters, and finally, just before what should have been my graduation, I managed to flunk Controls, a course required for my engineering degree.

That meant summer school: one chance at redemption. In most of India, summer means the monsoon season. In Kanpur, it's wetter than it is hot—and it is as hot as anywhere I've ever lived. The dormitories weren't air-conditioned, but the rooms did have French doors that opened onto a small terrace, and so, like every other sucker lucky enough to be taking remedial summer courses, I pushed my bed close to the French doors, so that at least my upper half could enjoy the occasional breeze. My worldly goods—my trunk of clothing, books, a stash of sweets I'd brought from home—went at my feet, at the other end of the room.

One morning, I woke up at eight or so, bleary eyed. Something was amiss.

Or rather, something was looking at me. Something with an impressive set of teeth.

I don't know if you've ever experienced the rapid transition from drowsiness to abject terror, but I don't recommend it. Every part of my body froze except my eyeballs, which swiveled on their bearings to take in my visitor. Staring back, fangs bared, was a rhesus macaque.

Too late, I realized my error: I had situated myself between my stash of sweets and any passing hungry simians.

Monkeys, in my opinion, are cute only on the other side of a television screen. They can be mischievous in all the worst ways, even violent, and they sometimes carry rabies. We stayed like that, face-to-face, for what felt like minutes. I pictured my obituary: *Once-Promising Student Mauled at Summer School.*

Five years earlier, in high school, no one who knew me would have predicted such an ignominious fate. Together with about 70,000 of my peers—the comparable number today is on the order of a million—I had sat for the Indian Institutes of Technology's entrance exam. It was, and remains, perhaps the most competitive exam in the world. With some variation by year, only the top 2 percent of test takers are accepted to any of IIT's campuses, which means for every fifty would-be dragon slayers, only one prevails. When my cohort's scores came in, I learned that I had placed among the top five hundred students in the country. The news came as a major relief. Though I had won the neonatal lottery in the sense that I had been born to educated parents who prized learning and independent thought, we were far from wealthy, and they had made it clear to me that if I were ever going to make something of myself, academic excellence would be non-negotiable. After I saw my name on that board, it seemed my adventure was off to a promising start.

But at some point at university, the going became a slog. It became difficult to figure out how and why the abstract coursework I was supposed to be learning mattered in the larger scheme of things. It wasn't my professors' fault; I was being taught by some of the better professors in India, and plenty of other students were doing just fine. And it wasn't that I didn't *want* to care about the material. I wished desperately, in fact, for the stuff I was being taught to fit into my head as effortlessly as it had in my childhood.

But instead, for the first time in my life, learning had become difficult. In a way, I almost envied the monkey slavering at the end of my bed. All he had to do was wander along the row of dormitory windows and take whatever food was on offer. To him, my university was a buffet, filled with treats just waiting to be discovered. It should have been that way for me as well. I could see some of my friends partaking in the spread of knowledge, but I couldn't force myself to eat. In a very real sense, the monkey was better adapted to my university than I was.

—

Today, more than thirty years later, I'm happy to report I survived the most dangerous part of my educational adventure. The macaque at the end of my bed scampered away, permitting me to spend the rest of the summer forcing down the curriculum of my Controls course (and reading J. R. R. Tolkien's entire *Lord of the Rings* cycle). If I hadn't been given that second crack at that course, however, or if something else should have come between me and my studies that summer, I shudder to think about the later educational journeys I would have missed out on.

Instead, today I find myself in a unique position: not just engaged in my own lifelong pursuit of knowledge, but also the lead guide for many others. I head the Open Learning effort at the Massachusetts Institute of Technology, where it's my job to fling open the doors of the MIT educational experience and extend aspects of it to as many people as might conceivably benefit.

My quest is hardly unprecedented. Many others across time and space have made similarly ambitious attempts: geniuses, visionaries, disciplinarians, anti-disciplinarians, administrators, philosophers, gurus, *literal saints.** Some failed, some succeeded with certain types

* Such as Saint Jean-Baptiste de La Salle, who dedicated his life to educating poor children in France and, in 1685, established what was likely Europe's first teacher-training school.

of students in certain conditions, and some saw their ideas spread nearly worldwide, only to recede. And so it would not be unreasonable to wonder what I, and the groups that I represent, bring to the equation that is so new and different. It can't be just my experience as an educator, although, if I may be so immodest, I am a good one. And it can't come down just to the transformative new education technologies being created at MIT and elsewhere. After all, would-be reformers have overhyped education tech for well over a century. To William Rainey Harper, who taught Hebrew at Yale University in the late 1800s, the transformative technology was the United States Postal Service, which he claimed would facilitate correspondence coursework "greater in amount than that done in the classrooms of our academies and colleges." In 1913, Thomas Edison voiced similarly high hopes for the motion picture, which he thought would render textbooks "obsolete." Next came radio, which was supposed to become "as common in the classroom as the blackboard," then television, which 1950s-era technophiles referred to as "the 21-inch classroom." In 1961, *Popular Science* predicted that by 1965, half of students would rely on automatic "teaching machines" for their instructional needs. By the 1970s and 1980s, computers had become the new top contender. And indeed, although they did soon become "a significant part of every child's life," as the MIT educational technology pioneer Seymour Papert foresaw in 1980, they have yet to live up to his 1984 prediction that "the computer will blow up the school." Far from it: Despite the onrush of technological changes that have come to education since the middle of the nineteenth century, most of us still teach and learn in classrooms that remain remarkably similar to those of 150 years ago.

And so, when I ponder what education could become, I find my thoughts wandering less to hypothetical, technological futures than to my own past: to my university dormitory and that monkey, for whom the physical architecture of standardized education spelled nothing but opportunity. In the course of my own journey, I would come to realize that those same sorts of educational structures—encoded not just in architecture but also in software, laws of the land, traditions, organizational rules, and unspoken norms—had ceased to function as a playground for my mind, and had begun to constrain my ability to learn. Many others had it worse. My most critical moment of distress had arrived late enough in my education that, with plenty of family support and few responsibilities of my own, I'd managed to muddle through. I had been one of the lucky ones.

ENTER THE EDUCATIONAL WINNOWER

I'm hardly the first observer to note that standardized educational structures, ostensibly put in place to nurture learning, can in fact impinge on it. In fact, a surprisingly wide variety of education reformers find common ground not just in their broad antipathy to the educational status quo, but also in how they describe it—which is to say, by analogizing schools to factories. As Alvin Toffler wryly noted in his 1970 book *Future Shock*, "the whole idea of assembling masses of students (raw material) to be processed by teachers (workers) in a centrally located school (factory) was a stroke of industrial genius." Today, you can find reformers of virtually all stripes inveighing against the so-called factory model, from the free-marketeers at the Clayton Christensen Institute ("The factory-model system" processes "students in batches") to the left-leaning Century Foundation ("The 'factory' model . . . tends to alienate teachers on the front lines"). Centrists, such as the Learning Policy Institute, have echoed the metaphor, and so has the arch-libertarian John Taylor Gatto, who describes school as "a kind of halfway house" preparing youngsters for "service to a mind-destroying machine." Tech evangelists, too, have hopped on the bandwagon: As Salman Khan, an MIT alumnus who created the Khan Academy educational video series, has said, "There is no need to continue the factory model inherited from 19th-century Prussia."

To the credit of all these critics, the standardization of school does indeed come with serious drawbacks. And yet, for all its prevalence, I still don't think the factory metaphor provides the best illustration of what mass-education-as-we-know-it does to learners.

For one thing, education reformers have been leveling similar complaints for a very long time. As the historian Sherman Dorn has pointed out, even as early as the decades leading up to the Civil War, American schools were replete with many of the supposed sins of today's "factory model," including mass-produced textbooks and a tradition of learning by rote. A wide wave of reforms took place, and yet, by the tail end of the nineteenth century, a new generation of reformers was already issuing complaints remarkably similar to those of their predecessors. This process repeated itself again and again, in the Progressive Era and throughout the twentieth century. In many respects, then, to rail against "factory-style" schools today is merely to slap a fresh metaphor onto an age-old struggle against repetitive, perfunctory instruction.

But still, it's easy to see why the factory comparison is so tempting. School, at a glance, appears to take in raw materials—human beings, in all their variety—only to produce a sea of similar-seeming graduates. It would only be reasonable to assume, then, that a shaping, molding process must be taking place between matriculation and graduation: clay, formed into dinner plates; gold, cast into bars; trees, whittled down to toothpicks.

Nowhere is the resulting homogeneity more apparent than in college admissions. In recent years, elite colleges in particular have found themselves with the strange problem of differentiating among students who, on paper, are essentially clones—at least in such still-important terms as their SAT and ACT scores and grade point averages. The product of the supposed "factory," at least when viewed through its preferred lens of standardized metrics, is undeniably consistent.

But molding isn't the only process capable of churning out such homogeneity. To the extent that the output of educational systems around the world differs from the input, it seems to me that a different procedure is just as responsible, perhaps more:

A *winnowing*.

By blowing air through a column of crushed seeds, a winnower can separate seeds from their shells, or grain from chaff. Small winnowers can take up as little space as a vacuum cleaner and can be kept in the corner of a bakery or coffee roasting shop. Large ones can be the size of a warehouse: an entire factory devoted to producing near-identical grains on an industrial scale not by molding or shaping, but by eliminating deviance. When building such a device, the question is always what level of error is acceptable. How much good grain is worth throwing out in order to achieve next to no chaff in the final product?

It's hard to pinpoint precisely how much raw human potential the global educational winnower routinely sacrifices for the sake of a consistent product, but there's every reason to believe the wastage is vast: a world's worth of attrition parceled out most visibly in rejection letters and underwhelming test scores, but also in less obvious forms: courses never taken, applications never sent, examinations never sat for, books never read.

Once you realize how education systems are set up not just to nurture, but also to cull, you begin to see it everywhere. We winnow in how we test, and we winnow in how we teach.

We also winnow, for that matter, in *who* we teach—and where, and when, and for how much. Take the widest possible view and try to imagine everyone in the world who might want to learn something via formal educational channels. Right from the start, a whole slew of access-related factors cut short educational journeys before they even begin. Maybe you'd like to take some higher-ed courses, but you live too far away from a college or university, or maybe you're "too old" to go to school. (No such thing, but on the other hand, holding down a day job certainly makes it harder to get to classes during the workweek.) Maybe you're a stressed-out parent, or someone with intensive eldercare responsibilities. Maybe you live in a region of the world without great schools, or your local schools won't admit people of the "wrong" race or caste or social position. In some countries girls are denied full access to education; in others, girls seem to zoom ahead in grade school only to be stymied later in their journey, or shunted toward stereotypically "feminine" fields while the boys get to play with robots. You might not belong to a family that expects you to go to college, or you might not belong to a community where higher education is the norm. As the economists Caroline Hoxby and Christopher Avery demonstrated in a 2012 study, a large number—"probably the vast majority," they write—of low-income, high-achieving American students simply never apply to a selective college or university, despite the increasing availability of generous financial aid packages.

And because not everyone is granted generous financial aid, the cost of education itself only adds capriciousness to an already ruthless winnower. Since the early 1980s, the full "sticker price" of college tuition has increased year in, year out, at more than double the rate of overall inflation. There are a number of reasons why, including the substantial administrative and physical-plant costs posed by major research efforts, declining government funding, and the economic fact that as salaries in high-tech fields rise, "high-touch" fields like teaching and healthcare must keep pace if they hope to stay in business.

But perhaps the most straightforward explanation for tuition's rise is the most persuasive: Simply put, the average undergraduate degree is still well worth the cost. As the MIT economist David Autor demonstrated in a 2014 paper in *Science*, pursuing a college degree remains one of the smartest financial decisions you can make, leading to a median lifetime benefit of roughly $500,000. One slightly jaundiced way to look at the mismatch between college's cost and value, then, is to recognize that the sticker price could, if anything, be worse, since the market would largely bear it. Indeed, one intriguing, model-based analysis, created by the economists Grey Gordon and Aaron Hedlund, supports this idea. They argue that broad, continuing student demand may be the *primary* driver of tuition increases: Because more students than ever want to go to desirable colleges and reap the college wage premium, while the number of seats at said colleges has remained largely stagnant, the price has gone up. (Meanwhile, at lower-tier colleges whose degrees don't confer much of a wage premium, you'd expect to see declining enrollment—and that's precisely what's happened in recent years, a trend that has hit for-profit colleges particularly hard.)

Despite the rise of needs-blind admissions and the best efforts of colleges and universities to provide significant financial aid, the cost factor still winnows away lower-income students with callous efficiency. The relationship between family income and college attendance is almost absurdly strong, as a team led by Harvard's Raj Chetty showed in 2014, with 25 percent attendance at the lowest income rung rising straight as an arrow up to 95 percent attendance at the highest income level. Worse, the lower your family's income, the less selective the institution you're likely to attend—a consequence not of differences in aptitude but rather of such concerns as the need to be close to home to help family members, or difficulty jumping through all the hoops required to put together a top-tier application package. Although in raw terms, less-selective colleges account for more of the U.S.'s income mobility than elite schools, elite schools have the edge on a per-student basis, and tend to offer better financial aid and support resources to boot. "Ironically," Stanford's Hoxby has said, lower-income students "are often paying more to go to a nonselective four-year college or even a community college than they would pay to go to the most selective, most resource-rich institutions."

Any plan to expand the learning horizons of people everywhere must involve recalibrating the educational winnower to be more inclusive, so that we stop turning away learners for such access-related reasons as income, geography, and timing. Necessarily, such a plan would involve adding "seats," be they physical or virtual, to the world's top-notch classrooms, for which demand is already high and growing. One force at odds with adding seats, however, is the all-too-prevalent idea that for a school to be top-notch, it must be "selective"—meaning it turns away most of its applicants. And therein lies a conundrum.

—

In fact, selectivity itself, and how we go about enforcing it, represents one of the more powerful, pernicious ways that our educational winnower squanders human potential. Slotting the best students into highly sought-after seats is, in its way, an admirable goal—after all, we should make the most of our limited resources. But the way we pursue that goal is built on a set of glaring assumptions: that our potential as learners is both knowable and fixed, more or less for life, perhaps starting as early as birth. These ideas, though highly debatable, comport with how most of us think about intelligence: If you're smart now, you'll still be smart in ten or twenty years. Or, as Lewis Terman, the father of American intelligence testing, wrote in 1919, "the dull remain dull, the average remain average, and the superior remain superior."

The *original* inventor of the intelligence test, the French psychologist Alfred Binet, disputed this sentiment. Binet was convinced of intelligence's fungibility. In fact, the whole point of his test, which he invented in 1905, was to identify members of the population whose intelligence might benefit from intervention. Soon, however, his test was coopted by Terman and like-minded thinkers, who approached the topic with a worldview steeped in the deeply unfortunate tradition now known as "scientific racism." Such scientists, in constant search of evidence that could prop up their prejudices, considered it a given that human intelligence was "chiefly a matter of original endowment," as Terman put it. "We must protest and react against this brutal pessimism," Binet lamented in 1909, to no avail. Within a decade, such pessimism was quite literally on the march. IQ tests, given to 1.7 million U.S. Army servicemen in World War I, led to the development of a National Intelligence Test for schoolchildren. By the 1930s, schools were testing for IQ and aptitude as a matter of course. They often started young; Terman insisted that "the limits of a child's educability can be fairly accurately predicted by means of mental tests given in the first year" of school, and "accurately enough for all practical purposes by the child's fifth or sixth year." Schools used the results to sort students into tracks: the high road being the "college track" or "academic track," as opposed to lower "general" or "vocational" tracks. Upper-track kids applying to college then found themselves facing yet another cousin of the wartime testing program: the Scholastic Aptitude Test, known today as the SAT, which began classifying students according to their supposed aptitude in 1926.

To give aptitude tests like the historical SAT their due, they can and do deliver a subset of learners from the jaws of the winnower. For generations now, by establishing a point of comparison across high school students, they have helped make admissions less contingent on personal connections (even if that transformation remains incomplete, given the extra consideration commonly granted to, say, legacy applicants). A good test score, meanwhile, can encourage students to give college a shot who otherwise might not. (Indeed, when the state of Maine began requiring students to take the SAT in 2006, 10 percent of students who would have otherwise skipped the test ended up attending a four-year college.) And, despite scientific racism's formative role in aptitude testing's history, a well-aimed aptitude test can actually expose racial bias in schools. In 2005, for instance, when Florida's Broward County introduced a universal, aptitude-based screening program to fill its third-grade "gifted" classrooms, instead of relying on parents' and teachers' nominations as in years prior, the Black and Hispanic populations of those "gifted" classes tripled.

But there are also problems galore to be found with intelligence and aptitude testing. It's impossible to gauge intelligence directly, so IQ tests attempt to sketch its outline by sending test takers through a battery of challenges, the combined results of which supposedly reflect one's cross-cutting intellectual chops. Some psychologists doubt that there really is a deep, generalized factor undergirding performance on all these sorts of subtests. IQ skeptics also like to point out that intelligence testing almost inevitably comes with blind spots and biases. Starkly drawn test questions fail to assess such virtues as creativity and interpersonal skills, for instance—and then there's also a long history of test makers coming up with questions that favor test takers of higher socioeconomic status (for example, "Define *regatta*").

But let's suppose, for argument's sake, that these caveats don't matter much: that IQ scores really do paint an accurate picture of one's ability to learn. Even then, the predictive power of IQ scores still comes up short, because IQ is not, as Terman supposed, fixed for life. Rather, Binet was right: IQ is alterable, fungible, contingent on your surroundings and experiences. Sorting students by IQ-style tests, then, can be a recipe for winnowing out students for their environmental circumstances, not their intrinsic aptitude. In fact, speaking in aggregate, the lower a family's socioeconomic status, the more variable (and therefore less heritable) IQ tends to be within that family—which implies that environmental factors wield particularly disproportionate influence over IQ scores in lower-status populations. Such external factors can be either helpful or harmful. Pollution, for instance, such as lead in drinking water or in the air, can take a lasting toll on IQ, as can childhood malnutrition, as well as childhood abuse and neglect. More temporarily, lack of sleep and acute stress can take a severe bite out of the cognitive processes required for good performance on IQ and other standardized tests. On the bright side, education can boost IQ: On average, every additional year of schooling you complete will garner you between 2.7 and 4.5 IQ points.

So malleable is intelligence, in fact, that the simple act of teaching students that it can improve—leading them toward what the psychologist Carol Dweck calls a "growth mindset"—can, in specific circumstances, cause noteworthy achievement gains. Perhaps the most obvious sign, however, that tests of intelligence and aptitude measure something fluid, not fixed, is the sheer scale of the test-prep industry that crops up whenever the stakes of such exams are sufficiently high. As the education historian Carl Kaestle has written, "Generations of affluent people buying test preparation to improve their children's 'aptitude' would prove the naïveté of calling the SAT a measure of 'aptitude.'" In the summer of 2019, in an attempt to mitigate such distortionary factors, the SAT-issuing College Board briefly announced that it would begin contextualizing its scores with information about test takers' neighborhoods and schools of origin, before walking the plans back a few months later. The tentative step would never have been necessary had SAT scores adequately represented aptitude in the first place.

Despite the continued influence of the SAT in college admissions, starting in the 1980s the American education establishment began to lurch away from aptitude exams and toward subject-matter-specific "achievement" exams, which test not for intelligence but rather acquired knowledge and skills. As these customs changed, a wave of bowdlerization washed over aptitude's remaining edifices. In 1993, the College Board changed the name of the SAT from "Aptitude Test" to the redundant "Assessment Test," and then, in 1997, did away with the words behind the letters entirely; SAT now stands for nothing in particular. Meanwhile, high schools' "college" and "general" tracks have slipped into a suit of subtler terminology: Advanced Placement and "honors" courses now form the rails of a college track that exists in all but name.

I wish I could reassure you that backward-facing achievement exams winnow away learners far less arbitrarily than forward-looking aptitude tests, but it's not clear they do. Acute stress, for instance, impairs cognitive processes that are indispensable for both sorts of exams. And then there's the matter of stereotype threat: the well-supported theory that negative stereotypes can provoke unfair distractions, doubts, and anxiety in students belonging to disadvantaged groups, harming their performance on high-stakes tests. Take, for instance, a group of boys and girls who, in 2013, posted identical scores on the Specialized High School Admissions Test, the sole criterion for entrance to eight of New York City's most selective public high schools. The following year, the girls went on to earn significantly higher GPAs than the boys, a sign that something about the entrance exam had disproportionately depressed the girls' scores. Perhaps the issue was that they were less willing to guess at answers than the boys, but it was just as likely that stereotype threat had erected an asymmetric distraction during the test—occupying girls' cognitive resources at precisely the moment when they needed them most.

—

If a learner's personal circumstances act like a thumb on the scale determining whether she's worthy of educational investment, and the biased methods used to sort students act like another, then there is, improbably enough, a third thumb (or perhaps a big toe) at work as well. It has to do with how we expect students to learn.

In even our best-intentioned attempts to shovel information into students' heads, educators routinely violate principles found throughout the research disciplines collectively known as cognitive science, from higher-order psychology down even to the molecular level. Every one of these violations—often the unintended consequence of scholastic norms and structures—comes at the cost of learning adventures delayed, diverted, or cut short.

Put simply: In our efforts to standardize education, we've made learning too damn hard. This unfortunate state of affairs harms students across the board, and may be especially fraught for disadvantaged groups who arrive in classrooms already stressed, especially vulnerable to the worst effects of biologically and psychologically unsound instruction. One powerful way to save vast swathes of the world's unduly winnowed learners, then, may be as simple as making learning more user-friendly, by identifying and eliminating unnecessary cognitive fetters.

Happily, I'm here to report that there are many feasible ways to do precisely this—not by watering down the knowledge and skills we hope to impart, but rather by deliberately rethinking how we teach, and how students might strive to learn. Such an approach would certainly not resolve deep, systemic biases, but nevertheless it stands to mitigate some of the excesses of our overzealous winnower.

Tailoring education to meet the specific, often surprising demands of the equipment between our ears will permit us to *increase* the depth, breadth, and rigor of learning—for a far wider population of students than currently benefits.

TO TEACH—OR TO SORT?

It's fitting that the modern science of learning may help pull us out of our current predicament, because its forebears certainly helped shove us into it. During the first few decades of the twentieth century, many of the emblematic features of contemporary mass education entered into widespread use while others, predating that era, were put to work in newly standardized ways. Abetting and inspiring this infrastructural sprawl was the young field of experimental psychology, which had begun to draw sharp, quantifiable edges around the once-gauzy phenomenon of learning.

In 1898, the experimental psychologist E. L. Thorndike declared to the world that learning came down to principles of mental association. He argued that when people—or, in the case of his original experimental subjects, hungry cats—do something that results in a desirable outcome, they become increasingly likely to repeat that action. In such instances, he maintained, a mental association forms: between a lever on the door of a cat's cage, say, and a tasty meal outside. As I'll explore in chapters ahead, using such fundamental building blocks, it's possible to posit an entire, freestanding theory of how human learning works, including in the classroom. (Instead of "lever" and "cat food," for instance, students might associate "5 times 7" with "35.") By chopping up learning into manageable, measurable chunks, Thorndike's theory created a scientific rationale for the standardization of schools, and his theories came with certain other attendant assumptions as well—about students. His model was, in essence, a new take on the old notion of the mind as blank slate, shaped by experience as opposed to innate predispositions. Thorndike's version, however, came with a near-Orwellian twist: Although all slates were born blank, some appeared to be more blank than others—or at least better able to absorb and parse incoming information. Like his contemporaries in the IQ-testing game, he believed this capacity to be both genetically determined and set for life, an idea that added considerable heft to the twentieth century's obsession with testing and sorting students. (Indeed, Thorndike and Terman worked together on both the Army's IQ testing program and the follow-up National Intelligence Tests.)

And so, even while some of his contemporaries were arguing that the purpose of education was to improve intelligence, Thorndike saw things differently. "The one thing that the schools or any other educational forces can do least," he told his students at Columbia University, was develop "powers and capacities." In a world where intelligence wasn't improvable, he recommended that schools focus their limited resources on those students best able to make the most of what they were given. It was imperative that schools winnow the chaff from the wheat.

It's hard to overstate the degree to which such assumptions became woven into the fabric of mass education in the United States, and much of the rest of the world, as the twentieth century wore on. Even today, many schools begin sorting students as early as the end of elementary school, as though their promise as learners were both set for life and entirely knowable. (It's neither.) We attempt to fill them with facts as though their heads are an empty bucket. (The better metaphor, as I'll explain later, would be a growing tree.) We spur them to learn with carrots and sticks like GPAs and test scores. (Instead, we could focus on building natural interest and curiosity.) Finally, at the end of this process, a very specific sort of student is left standing—someone who is in no way undeserving of further educational investment, but who is also in no way representative of the full scope of human potential that went in.

Even as mass-education structures crystallized around Thorndike's point of view, however, the field of cognitive science continued to advance, creating a gap between the leading edge of scientific knowledge and the increasingly outdated assumptions frozen at the core of standardized education. The issue wasn't that Thorndike was wrong. Rather, his theory of learning turned out to be reductive: an oversimplification of what we now understand to be an extremely complex process. In fact, the learning mechanism that Thorndike theorized bears more than a passing resemblance to today's most commonly accepted neuroscience model for how information is represented among the brain's synapses—a remarkably prescient finding. What he failed to anticipate, however, was that his model might compose only the ground floor of what many cognitive scientists now think of as a high-rise of sorts, with individual neurons and their synapses supporting the activity of specialized regions of the brain; the physiological brain supporting psychological processes; physio- and psychological processes both contributing to what we think of as the conscious mind; and above that, multiple minds interacting with each other in classrooms and beyond. At every one of these levels of organization, researchers have uncovered processes that turn out to be absolutely critical to successful learning—and which, when interfered with, can quickly bring an educational adventure to a grinding halt.

Many of our inherited education practices, with their hundred-plus-year-old underpinnings, sit uneasily with these newer findings. If my mission to foster learning far and wide is to have a prayer of a chance, it must avoid replicating these same old mistakes on a larger scale. Rather, we must first undertake a cognitive reckoning of sorts. Based on what we now know about the levels of the learning brain and mind, what—as teachers and learners both—can we do to make the act of learning more user-friendly?

—

Perhaps the most obvious impediment to making learning easy is the pervasive notion that learning *should* be difficult. This idea is as prevalent at MIT as anywhere—perhaps more so. As an informal slogan, MIT's undergraduates have adopted the initialism "IHTFP"—short for "I Hate This F***ing Place," intoned usually (but not always) with a rueful smile—which gives you one good benchmark of how taxing they find their coursework. Challenging students to develop rigorous, deep, and readily activated bodies of knowledge is essential. Imposing onto this already difficult undertaking additional, *unnecessary* hardship, however, is both cruel and arbitrary—like requiring Olympic sprinters to first qualify in a karaoke competition.

And yet, another of MIT's cherished tropes hints at precisely this sort of needless difficulty. It's long been a point of masochistic pride that learning at MIT is like drinking from a high-pressure fire hose. Our students, this bit of campus lore suggests, are smart and tough enough to gulp down a flow of information capable of stripping paint. And so, in 1991, when a group of pranksters set up a water fountain on campus that literally dispensed drinking water from the tip of a fire hose, MIT made it into a permanent exhibit. The fire hose fountain now graces the entrance to the Ray and Maria Stata Center, the building where most of MIT's computer scientists live, with a commemorative plaque explaining about high-pressure learning. Every time I walk past it, I grimace, because the sculpture suggests two misguided ideas. First, that apparently learning is a difficult endeavor at MIT. That's strange, considering that as the world's premier institute for technical education, blessed with an utterly fantastic student body, you'd think we would find it easier to teach. (And in fact, we are *far* better at teaching than the fire-hose metaphor suggests.) And second, it implies that people who, for whatever reason, are unable or don't care to sip from the stream of pressurized truth don't belong at MIT. Indeed, perhaps advanced education is not for them. Perhaps *knowledge itself*, beyond what they catch as it flies by in the news and on social media, is not for them.

Or more likely: We've been thinking about learning all wrong. We've internalized the idea that learning is meant to be an ordeal through which students must persevere or fail. I'd like to step back and ask why. Human beings are built to learn. It would not be too overdramatic, in fact, to argue that our learning ability is our birthright as a species, hard won through eons of natural selection. The difficult part should be over. When did the abstraction of useful knowledge from the world around us become attainable only through bitter perseverance? When did learning become not the fun sort of adventure story, but a grim slog against the odds?

In an education system set up in part to sort the supposed elect from the unworthy, any unnecessary impediment to the act of learning itself will inevitably push a subset of students toward the latter category: a flunked exam here, a denied college application there, diminished educational prospects overall. More troublingly, there's every reason to believe that that subset—the winnowed—will disproportionately consist of students who arrive at school pre-burdened with the sorts of social and economic disadvantages I've already discussed. One obvious, worthy response to such problems is simply to fix them: to lead a full-frontal attack on poverty and racism and sexism and all the other things that keep good people down. In fact, if we ever hope to live in a more just society, such a direct approach is simply non-negotiable.

But even in a world miraculously without such obstacles, learners would still be hamstrung by cognitively clumsy instruction. To truly make education as equitable as possible, it is imperative that we shed the myth that serious learning must be difficult, and find ways to make instruction far more cognitively user-friendly. Such an approach would save students from undue winnowing while helping others achieve new heights of academic excellence.

And if we can save a few students from the ruthless machinery of the educational winnower, then perhaps we can also multiply that effect, taking our new and improved approaches and making them accessible to a wider world of learners. This is where new instructional technologies come in. As the peddlers of earlier technologies discovered the hard way, a hot new instruction medium alone—even one capable of reaching vast numbers of students—isn't likely to drastically change how most people learn. The idea of using educational technologies as a vehicle for a more brain-friendly mode of education, though promising, isn't exactly new, either. But as I explore in the pages ahead, both our understanding of how learning works and the capabilities of new instructional media have reached a point where their combined potential raises possibilities too promising to ignore.

Let me say this before we go any further: I believe the best education is still a human-to-human education, and I don't see that changing anytime soon. We're only now, however, starting to see what great human teachers can accomplish while wielding truly powerful education technologies—especially when given the freedom to act outside the inherited, often procrustean structures of traditional mass education.

The need for more user-friendly, accessible learning has never been more urgent. Admittedly, in saying this, I'm guilty of the long-standing, unfair practice of dropping off the world's problems at the classroom door for educators to solve. But the fact is, the current moment does present a suite of issues that demand uniquely *educational* solutions. From the perspective of learners, the pace of technological change is altering the rules of work, even as careers are growing longer than ever. We need better, time-and-energy-efficient ways for busy midcareer workers to keep current or change tracks as needed. Meanwhile, at a time when intergenerational income mobility has been stagnant for decades, we need new ways to bolster one of the few forces capable of helping individuals vault free of their parents' circumstances: a college education.

And then there are the broader problems faced by entire societies: climate change perhaps most existentially, but also such concerns as pandemic disease, the medical and caregiving costs facing countries with large aging populations, income and wealth inequality, and finding ways to reattach gains in living standards to economic growth, among many others. Unfortunately, we're devoting only a small portion of our collective brainpower to the sort of innovation we'll need to solve such pressing issues. Only a small fraction of people is ever even given the *chance* to innovate. In an attempt to probe who gets to participate in technological innovation, Raj Chetty of Harvard and his team have shown that the vast majority of patent-holding inventors fit an alarmingly narrow profile: mainly boys from wealthy families and rich neighborhoods. "These findings suggest that there are many 'lost Einsteins,'" they write, "individuals who would have had highly impactful inventions had they been exposed to innovation in childhood—especially among women, minorities, and children from low-income families." This innovation imbalance equates to less income for those underrepresented groups and less innovative energy devoted to solving the unique problems they experience. Just as alarming, it means that in terms of the gigantic, urgent challenges facing societies around the world, we're fighting with one hand tied behind our back.

Scientists, meanwhile, are making their most groundbreaking discoveries later and later in their careers, likely because in any growing area of research, it takes each generation longer than the last to get up to speed. We need ways to hasten students' journey to the cutting edge of research, and we need people from more diverse backgrounds taking part.

One answer to both problems is to make learning easier: more user-friendly and far more accessible, for all different sorts of people, all around the world.

One way or another, learning is going to be an adventure. It can be a story of Herculean perseverance or it can be a voyage of joyous discovery. By erring on the side of the fun kind of adventure, we may help more people learn, and realize more human potential, than we've ever seen.

My lucky position at MIT has given me insight into how this could happen, and now it's my pleasure to share it with you. In the chapters ahead, you will meet the scientists, educators, and engineers who are changing the way we're thinking about learning, on MIT's own campus and beyond. If learning is an adventure, then learning *about* learning may turn out to be one for the storybooks.

Part One

LEARNING IS SCIENCE AND SCIENCE IS LEARNING

- I -

THE LEARNING DIVIDE

It was the last day of February 2017, and Amos Winter, an assistant professor of mechanical engineering at MIT, was warning the group of sophomores in his afternoon lab section about the destructive potential of their batteries. Though supposedly safe, in the unlikely event of a sudden discharge, each of the lithium polymer batteries scattered on the conference table possessed enough energy to maim, even kill.

How much energy, exactly? "Go ahead—slam it into a calculator," he said. After approximately ten seconds, anyone who had worked it out was keeping the answer to herself, so Winter bounded over to a whiteboard. You know the capacity of the battery, he explained, which came labeled in units of milliampere hours. "You basically just add in time to figure out energy in joules," he said, and in short order, the answer was on the board: 13,320 joules. "That's the equivalent to lifting a Honda Civic ten meters off the ground," he said. "Imagine a Honda Civic falling on your hand"—that's the kind of damage an exploding lithium polymer battery could inflict. If the casing on such a battery begins to bubble, he said, chuck it in one of the lab's many sand buckets and run in the opposite direction.

In the absence of any such catastrophes, however, class would continue to hum along as it had for the first few weeks of the semester. In addition to the batteries, sitting on the table in front of each student was a simple robot—two wheels and a skid designed to drag along the ground—which would serve as a sort of training vehicle, in anticipation of the more complex robots the class would build later in the semester. On these practice bots, which Winter dubbed "Mini-Mes," the students would learn mechanical engineering principles ranging from simple to complex. They would start by learning to code a microcontroller (that is, a very small computer) to run an electric motor; later, they would instill in their Mini-Mes the capacity to navigate the world autonomously like rudimentary self-driving cars. Along the way, they would learn not just robotics knowledge and skills, but how to think like designers and engineers. They would come to understand how to approach a task creatively, to spot issues before they become serious problems, and, perhaps most important, to gain a level of trust in their own ability to guide a project from early phase, when there are innumerable paths to a desired solution, to late, when there's only one best way forward.

That was the learning progression in theory, at least. In practice, some of Course 2.007's students were coming to it with more engineering experience than others. Some had competed in high-school robotics tournaments. (The best-known extracurricular robotics organization, FIRST Robotics, had actually spun out of MIT's original version of Course 2.007, back in 1989.) And the rumor mill had already made it known that one student, Alex Hattori, had competed on BattleBots, a televised contest known for its metal-on-metal violence. He and his teammates had sent a buzz-saw-wielding robot the size of a manhole cover into a gladiatorial arena, to wage war on opponents with names like SawBlaze and Overhaul.

To the other 164 students who lacked such head starts, these advantages were cause for real concern. In MIT's charged academic atmosphere, stress among students is a perennial issue, and unnecessary competition, usually over grades, does not help. Most of the time, the Institute works hard to dampen this instinct—for instance, by abolishing grades in the first semester of freshman year. But Course 2.007 is different. Competition is baked into it at a deep level, and is the reason why it is arguably MIT's most famous undergraduate offering. At the end of every spring semester, the course culminates in a robotics showdown, which draws hundreds of spectators from across campus and beyond. The winner achieves lifelong bragging rights, entering MIT Valhalla while notching one heck of a résumé bullet point.

Brandon McKenzie's gaze slid to his lab mates seated around

the table. A varsity swimmer who had competed in the Division III national championship as a first-year and would return to the championship series later in the semester, he had thus far maintained a perfect 5.0 GPA despite spending eighteen-plus hours per week in the pool. He was not used to the sense of falling behind, and yet there was no shaking the feeling that others were several lengths ahead of him in the race to build serious, competition-worthy robots. He had come to 2.007 with next to no practical robotics experience, and there were a few others in the same predicament—Amy Fang, for instance, at the other end of the table, and Josh Graves, Brandon's roommate, teammate, and all-around co-conspirator, at his right elbow. But then there were folks like Jordan Malone, seated directly across from Brandon, whose computer-aided-design prowess Winter would later describe as a "super power." (And that wasn't even the most impressive thing about him: Although he never brought it up unbidden, everyone knew that Malone, a short track speed skater, had brought home Olympic medals from Vancouver and Sochi, prior to enrolling at MIT at age thirty.) And there was Zhiyi Liang—Z, for short—a joyful mad-scientist-in-training who seemed to come to class every week having produced a new mechanical marvel in his downtime. Brandon expressed no animosity toward his fellow students; indeed, he would become the lab's most reliable source of fist bumps and backslaps in the weeks to come. But then again, he didn't feel any animosity toward his swimming teammates either, and that certainly didn't stop him from trying to outswim them.

Winter doled out off-brand Arduinos: microcontrollers that would inform the movements of the class's Mini-Mes today and, later, their full-fledged, competition-ready robots. That morning's lecture had concerned the mechanics of brushed, direct-current motors, the simplest type of electric motor. Now, mere hours later, Winter was taking his students' understanding of DC motors as given and demonstrating how they could be put to work. As Winter blasted through a series of reasonably complicated concepts, Brandon scrambled to take in his words while also adjusting his Mini-Me's physical wiring and fiddling with his Arduino's code on his laptop. He sensed he was in danger of sliding even further behind.

"I felt a little discouraged," he said later. Although Arduino's programming language, C++, was basically new to him, some of his classmates seemed to know it "like the back of their hand." He was keeping up for the time being, but he knew that the moment his attention strayed he would find himself stranded. This course had a sink-or-swim quality to it that felt all too familiar. It was as though he'd been chucked into the deep end of a pool but didn't yet know how to stay afloat. And although there were plenty of instructors looking on, telling him *how* to keep his head above water, it was up to him to apply that information in a way that actually worked.

Provoking that conceptual shift—from theory to practice, from inert to activated knowledge—is what 2.007, at its core, is all about. The course assumed its modern form in 1970, when a young professor named Woodie Flowers took the reins. In the decades that followed, as the beloved professor became a professor emeritus, he took on local-celebrity status on campus, where he tended to pop up periodically, Stan Lee–like with his trademark mustache and silver ponytail, to speak about learning. Every time, he stressed a single, crucial point: the difference between "learning calculus" and "learning to think *using* calculus."

To educators like me, who hope to produce students capable not just of earning good grades but of exerting their knowledge in the wider world, this distinction is of the utmost importance. But for students accustomed to clean borders around their education—the four edges of an eight-by-eleven worksheet, four walls of a classroom, four-year progression through high school, then college—it can be unsettling to step out into the messiness of the real world, even temporarily. And so every year, while some of 2.007's students exult in the chance to get their hands dirty, others hang back, sometimes for weeks, to get their bearings.

Across the table from Brandon, Amy Fang had already sized up her classmates. Arriving at MIT, she later recounted, she found her fellow students so intimidating that she decided that staying in the middle of the pack was a perfectly worthy goal. "I try to be average," she said, laughing. For the moment, as she connected her servomotor to her Mini-Me's breadboard, any sort of noteworthy success in 2.007's semester-end competition seemed unlikely. A decent grade would be challenge enough: 2.007 had no exams, and success in the final competition itself would have no direct bearing on grades. But the design of the students' competition robots would eventually come under close scrutiny, and in the meantime the course's significant homework load felt like a millstone. It came in two parts: written homework—a weekly set of four tricky, multipart engineering problems—as well as a weekly physical challenge that students were charged to accomplish using their Mini-Mes. This week, they had to submit video of their Mini-Mes heading forward a preordained distance, performing a U-turn, returning, and coming to a complete stop—all without any input from a remote control. Later in the semester, some students' robots would pull off autonomous tasks of greater complexity, such as using a light sensor to follow a line drawn on the ground.

As Winter explained about this path-following strategy, Mo Eltahir, a student from nearby Lexington, Massachusetts, spoke up. "It's cool because every lecture is stuff you can think of to use for the competition."

Winter let out a laugh. "Would you go so far as to say you're learning?"

They were indeed. In the demanding weeks to come, they would etch new memories more deeply than they probably knew—and not just any old memories, either, but highly contextualized ones that would empower them to both understand their world and influence it. To encourage this process, Course 2.007 does away with many of the problems endemic to traditional classrooms and lecture halls, which have hardly budged over the course of more than a century, and replaces them with something better.

In 1902, the educational psychologist and philosopher John Dewey ("America's foremost philosopher of his time," his *New York Times* obituary would eventually read) laid out the "typical evils" of classrooms he'd observed. Such incorporeal subjects as mathematics, separated from the objects and processes that numbers represent, and geography, divorced from geological and historical events, lacked "any organic connection with what the child has already seen and felt and loved," he wrote in his book, *The Child and the Curriculum.* As passive recipients of knowledge-made-inert, students went through the motions of school without ever feeling truly motivated to learn, mainly because the stuff they were being taught didn't relate to their day-to-day concerns and goals. Today, even when schools manage to better contextualize what they're teaching, the same central assumption still persists: Students are expected to learn for the benefit of their future, not present, selves.

Course 2.007 isn't like that. First, thanks to its sink-or-swim nature, students don't have the luxury of a lazy, memorization-based approach to studying. Rather, any theory taught in class can and must be immediately applied—after all, if you don't apply it, your opponents will. In part because students thrill to the experience of seeing their knowledge translate into real-world engineering powers, and in part because there's glory on the line, the course's ability to motivate learning is second to none. Even students with a long history of doing the bare minimum* wind up inspired despite themselves, spending far more time in the lab than is strictly necessary, screwdriver in hand, notebook computer open. In the space of a single semester, the course launches experienced hobbyists and the uninitiated alike toward professional mechanical engineerdom in a way that is sometimes imitated but never, in my completely biased opinion, exceeded.

Today, when I contemplate the task before me, my thoughts never stray far from 2.007. At minimum, any education scheme worth its salt must not only deliver knowledge, but do so in a way that is highly engaging—and then activate that knowledge, so its owner can do real work in the world. Course 2.007 vaults these bars with plenty of clearance.

The problem is a matter of access. Course 2.007 is extremely expensive to offer, costing MIT far more per student than each would pay in tuition, even without financial aid factored in. This owes in great part to the costs of keeping the laboratory up with the cutting edge of fabrication technologies while still safe for novices, which equates to a lot of highly trained personnel.

Pragmatically speaking, anyone hoping to provide an educational experience of this caliber to vast numbers of people either must be mind-bogglingly rich or else find a creative way to pull it off at scale. After all, you can't put a billion people through Course 2.007; even if you somehow found enough teachers and built enough laboratories,

* And there are students who fit this profile at MIT—they just tend to do the bare minimum needed to earn an A, as opposed to a B. I recognize the type; I used to be one.

according to one back-of-envelope calculation I've done, it would still cost more money than exists in the world.

It's certainly possible, however, to disseminate knowledge on that order of magnitude for far less: Wikipedia alone has done that, for instance, to great and deserved acclaim. So has MIT, in our own, smaller, way: Starting in 2001, through our OpenCourseWare initiative, we've made essentially all of our course materials freely available to anyone with an internet connection. But making information available is not the same as providing an education. And so you might reasonably ask: Can a scaled-up education scheme ever replicate what a skilled teacher in a traditional classroom can achieve, let alone the motivating, contextualizing effect of Course 2.007? Will its students ever truly make the jump from understanding calculus to *thinking using calculus*?

TEACHING MACHINES, TEACHING HUMANS

Would-be innovators have dreamed of distilling and mass-producing the secret sauce of education for well over a century. As early as 1912, the psychologist E. L. Thorndike, already well on his way to reshaping how America thought about learning, mused: "If, by a miracle of ingenuity, a book could be so arranged that only to him who had done what was directed on page one would page two become visible, and so on, much that now requires personal instruction could be managed in print." In 1953, the Harvard psychologist B. F. Skinner, in many respects Thorndike's intellectual descendant, attempted to realize this science-fictional notion by building a series of "teaching machines." One of them can still be found at Harvard, tucked away on the ninth floor of the university's William James Hall. The wooden, rectangular machine would have covered most of a student's desk while she worked at it, making her way through the series of questions printed around the edge of the paper disk inside. A small rectangular window in the machine's bronzed lid displayed a single question at a time as well as the answer to the previous question, and a nearby aperture let the student scrawl her answer longhand on a strip of paper tape that emerged briefly from the machine's innards before plunging back in. She would compare each written response to the correct answer and, by pulling a lever, mark herself correct or incorrect. (Teachers could check students' answer tapes for inconsistencies after the fact.) Once she'd answered every question on the disk, it would then spin more freely, stopping only to re-pose those questions she had initially gotten wrong, a process that would continue until she had answered every question correctly. Students would move along at their own pace, advancing from one disk to the next. The education revolution, Skinner believed, would thus be personalized. As one student memorably put it, "The eggheads don't get slowed up; the clods don't get showed up."

Just to the right of the window where students recorded their answers, the bronze surface of the teaching machine on display at Harvard is worn down to the underlying gunmetal: the result of a decade's worth of wrists rubbing the same spot. As the 1960s gave way to the 1970s, however, the hype, too, faded from teaching machines, and Skinner's invention vanished from both schools and the national conversation. They had proved less effective in the wild than in the lab, many in the public found them creepy, and much of the buzz surrounding the mechanical devices was finding its way over to electronic ones: computers and the nascent field of educational software. But still, neither the original teaching machine nor its computerized equivalents have ever transformed the classroom as promised.

And why not?

Perhaps most egregiously, teaching machines and their immediate electronic descendants were *boring*. Once the novelty wore off, many students said they hated the things—a major indictment of Skinner's entire project. Although it can feel momentarily gratifying to get answers right, and it's often good to move at your own pace, Skinner's approach proved overly simplistic: blind to larger concerns like motivation, contextualization, and social isolation. Like his predecessor E. L. Thorndike, Skinner was a reductionist, seeking to explain learning in terms of its most fundamental constituent parts. To his credit, reductionism done right is one of the most powerful conceptual tools available to scientists; it's what allows us to understand chemical processes in terms of atomic physics, for instance, or the concept of temperature in terms of the average kinetic energy of molecules. But "done right" assumes that the fundamental particles and processes you've isolated really do explain the workings of the larger system in question. When reductionism goes wrong, however—when scientists and engineers underestimate the complexity of a system—disaster can result: airplane crashes, stock market crashes, and everything in between.

Somewhere on the continuum between Skinner's reductive teaching machines at one extreme and the relatively holistic approach of Course 2.007 at the other, something important clearly gets lost. And that wouldn't be much of a problem—except for the fact that the structures of standardized education, guiding and confining students and teachers both, happen to be built of the same intellectual bricks as Skinner's teaching machines. In fact, I'd go so far as to claim that, to the extent that traditional classrooms today successfully help their students develop deep, contextualized, activated knowledge, it's due mainly to the skill of teachers working *despite* the limits imposed on their medium—many of which can be traced straight back to Skinner's intellectual godfather, E. L. Thorndike.

What's so striking, looking back, is how at the very same time that Thorndike and his allies began pursuing the research thread that would undergird and justify the form mass education took in the twentieth century, a separate research thrust, built on a wildly different scientific ethos, offered an alternate path—one far closer in spirit, as it turns out, to Course 2.007. In fact, although the story of education at the turn of the twentieth century features enough characters to rival the greatest hits of Russian realist fiction, the story of the science *behind* the education is a much tighter narrative, boiling down mainly to a contest between two figures in the nascent field of educational psychology. It did not end in a draw. As the education historian Ellen Condliffe Lagemann has written, "I have often argued to students, only in part to be perverse, that one cannot understand the history of education in the United States during the twentieth century unless one realizes that Edward L. Thorndike won and John Dewey lost."

Although the more obvious set of battle lines that formed between the two scholars concerned educational practice, their higher-level disagreement on how to conduct and apply science itself will hold the greater significance for our journey in the pages ahead. If we ever hope to usher in a vision of education at scale that is not reductive but rather *expansive*—closer to 2.007 than teaching machines—we must first plumb the scientific divide that separated Dewey and Thorndike.

DEWEY'S LABORATORY SCHOOL

At the very tail end of the nineteenth century, the young field of educational psychology was being pulled apart. Holding fast to one side were the defenders of *mental discipline*, an archaic theory that analogized the brain to a muscle in need of exercise, typically achieved via rote practice. The content to be learned mattered less than the effort involved in learning it, and so, strangely enough, the tradition found its strongest advocates in the classics-enamored educational old guard. Since you had to exert your mind on something, their thinking went, it might as well be something in Latin or Greek.

Leading the other side was one G. Stanley Hall, a psychologist known for applying newfangled experimental techniques to a worldview centered on Charles Darwin's theory of natural selection. In a plan that might sound downright reasonable to modern ears, Hall wanted schools to put away the mental barbells and structure their curricula instead around students' natural interests. There was a catch, however: "Interest," as Hall defined it, amounted to a peculiar mix of historical myth and biological fantasy. Far from alone among academics at the time, Hall believed (white) Western culture had achieved an advanced stage of development to which other cultures and races could only aspire. Many educational theorists (including the otherwise egalitarian Dewey) took this unfortunate idea a step further, tracing metaphorical connections between the development of children and the "development" of the world's cultures. Hall took it further still, believing in a hard, biological relationship between the supposed stages of human history and those of childhood.

The resulting pedagogical strategy was, in its way, nearly as rigid as mental discipline. Elementary school, to Hall, should consist mainly of play, stories, and the study of nature: the sort of activities ancient, "primitive" societies got up to. Only at age eight, he decreed, should reading and writing enter the picture, to correspond with the historical origins of written language. However, because "reason is only dawning" at that age, he wrote, teachers should introduce abstract ideas only through rote memorization, and wait until their students had surpassed the so-called Homeric Stage before explaining the *how* and *why* behind those concepts.

Encountering the absurd contest between the champions of "effort" and "interest," as the two sides were known, both Dewey and Thorndike sought to break it up. That wasn't all they had in common: Both turn-of-the-century education psychologists were heavily influenced by William James, widely considered the progenitor of American psychology; both taught as colleagues at Columbia University in the early 1900s; and they even looked somewhat alike, with dark, center-parted hair and impressive mustaches. But the impression of similarity lasts only that long. To linger on a superficial point, take their mustaches: Thorndike's triangular mustache looked as though he trimmed it with the aid of a protractor, while Dewey at one time sported a bushy soup strainer that, to a modern observer, might connote horses and six-shooters. Dewey's approach to educational research was just as unreservedly bushy, reliant on relatively naturalistic methods of observation as opposed to the stripped-down experimentalism that Thorndike wielded with surgical precision.

Dewey's debut in the American educational conversation took the form of a widely read monograph arguing that both sides of the effort-versus-interest dialogue were guilty of ignoring students' true interests. Might it be possible to begin the education process with children's "urgent impulses and habits," he wondered, and then nurture the growth of a larger body of knowledge around them?

That same year, Dewey opened the doors of the Laboratory School at the University of Chicago to find out. William Rainey Harper—the onetime professor of Hebrew at Yale who, if you recall, waxed bullish on correspondence courses—had been tapped to become the University of Chicago's first president. Intensely interested in pedagogy, he began assembling a dream team of educational researchers, including both Dewey and his soon-to-be collaborator, Francis Parker. The timing was perfect: Dewey, sick of all the "general theorizing" around effort and interest, had begun to mull a way of testing the various theories floating around the education world—most importantly, his own. A school, he decided, could function as a laboratory for the study of learning. "If philosophy is ever to be an experimental science, the construction of a school is its starting-point," he wrote. "The school is the one form of social life which is abstracted & under control."

—

The key word was *abstracted*. Today, one typically encounters the term in adjective form (an abstract oil painting), but Dewey often used it as a verb, to signify how, starting at a very young age, both learners and scientists actively strive to single out what is noteworthy about the world around them. From infancy, he reasoned, raw data wash over our senses, and to make head or tail of it all, we begin performing observations and even subtle experiments, like tiny scientists. "The native and unspoiled attitude of childhood, marked by ardent curiosity, fertile imagination, and love of experimental inquiry, is near, very near, to the attitude of the scientific mind," he wrote in 1910.

To Dewey, the distance between the sort of firsthand knowledge children gain through personal experience and the secondhand knowledge passed down generation to generation (algebra, which snakes are venomous, the location of Des Moines) was significant, but bridgeable. The educator's challenge was to close this gap: to help the child's interest stretch from items she had abstracted for herself to the *pre-abstracted* knowledge of her forebears—and thus, wrote Dewey, "short-circuit for the individual the slow progress of the race." Crucially, by starting with students' day-to-day concerns, this approach had the capacity to make learning feel immediately rewarding. It was a sharp break from the prevailing idea (then as now) that school should be endured "as a preparation," he wrote, "for something else, or for future life."

Well-planned experiments (including informal and thought experiments) designed to abstract theoretical truths from the concrete universe in all its messy, noisy glory, were the key to the entire operation. The Laboratory School was designed to enable such experimentation in several ways. Its students, natural-born experimentalists, would be encouraged in their instinctive endeavors. Its teachers would be given great leeway to experiment with day-to-day pedagogical strategies, based on how the students responded. And the school as a whole would serve as an experimental test subject:

to verify whether its approach could truly lead children from their immediate interests to the wide base of knowledge they would need in a rapidly changing world. All told, the school would bear "the same relation to the work in pedagogy that a laboratory bears to biology, physics, or chemistry," Dewey proclaimed.

—

The Laboratory School opened in January 1896, with a group of 16 students, including Dewey's own two children, and two teachers. By 1902, it would swell to accommodate 140 students, 23 teachers, and 10 graduate-student assistants. The school devoted no courses exclusively to reading, writing, or arithmetic. Rather, students were encouraged to treat such fundamental skills not as an end goal for learning, but rather as a means of solving whatever problems they encountered. Cooking, for instance, provided "a natural avenue of approach to simple but fundamental chemical facts and principles, and to the study of the plants which furnish articles of food," Dewey wrote. Gardening gave the students the opportunity to perform biological tests on the viability of seeds; smelting led to an explanation for why charcoal burns hotter than wood. Students made textiles from the ground up: they sheared sheep, carded and spun the wool, and then wove it into garments.

Such methods might today be called "interdisciplinary," but, as Dewey liked to point out, the academic disciplines are actually arbitrary constructs, none of which, he wrote, is "eternally set off" from any other. The school's approach, Dewey claimed, didn't just impart richly contextualized facts and skills, but also provided students "with the instruments of self-direction" and taught them how to work with others in their "miniature community, an embryonic society."

Dewey hoped the microcosmic community he was building at the Laboratory School would give rise to "a larger society which is worthy, lovely, and harmonious." To this very day, historians debate why this sunny vision never came to pass. It might be due to the fact that Dewey left the school suddenly in 1904, just eight years after its inception. Compared to the lofty heights of his ambitions, the events leading to his exit seem impossibly banal, involving a clash of personalities and charges of mismanagement against Alice, Dewey's wife, who was serving as principal. When Dewey left, he was snapped up almost immediately by Columbia University, where, starting in February 1905, he joined the philosophy department. Although in a sense his career as a philosopher of education (among other topics) was only beginning, his days conducting original education research were over, which put his ideas at a disadvantage in the decades to follow.

Another likely reason why his school plan failed to spread beyond Chicago is that his ideas simply ran counter to the zeitgeist, clashing against what society at the turn of the century was looking for from its schools. Especially at a moment when E. L. Thorndike, Dewey's intellectual rival—and, now, colleague—was beginning to offer an attractive alternative.

THE BASEMENT AND THE ATTIC

"I just cannot understand Dewey!" So proclaimed a frustrated Edward L. Thorndike about his Columbia University colleague. From the very start, despite having plenty in common, including faith in the ability of experiments to plumb the mysteries of the universe, Dewey and Thorndike built their respective worldviews around antithetical assumptions. So irreconcilable were their outlooks, in fact, that the pioneering psychological writings of William James served as a Rorschach test for the pair, drawing out their differences. When Dewey read James's articles in the 1880s (eventually collected in James's 1890 omnibus book, *The Principles of Psychology*), he saw clear support for a holistic approach to experimentation. As James's *Principles* made clear, the human mind had evolved in a social setting. Attempting to divorce the mind from that setting for the purpose of an experiment, Dewey determined, would be just as futile as studying plant growth in the absence of sunlight. A stand-alone human mind was simply too *small* to experiment upon by itself, so Dewey built an entire miniature community in order to have a system big enough to study.

A chance encounter with James's book would send Thorndike, meanwhile, hurtling in the opposite direction, toward reductionism: breaking systems down into their smallest component parts. To his point of view, the learning mind was too *large* to think about as a whole; insight would come only from isolating the smaller, simpler mechanisms that made it tick.

As an undergraduate at Wesleyan University, Thorndike entered an academic competition requiring him to read from James's opus, which he said he found "stimulating, more so than any book that I had read before." Perhaps most exciting were the two-hundred-odd pages concerned with James's experimental studies—an approach, Thorndike quickly realized, that reverberated with power. The old mental disciplinarian notion that memory constituted a "faculty" improvable through exercise, for instance, fell to pieces under James's scrutiny. James had memorized 158 lines of the long poem *Satyr*, by Victor Hugo, which, he recorded, took him about fifty seconds per line. Next, over the course of a month, he memorized the first book of *Paradise Lost*, which should have, in theory, strengthened his faculty of memory to peak human levels, making him the Usain Bolt of poem recitation. However, when he returned to *Satyr* and attempted to memorize another 158 lines, each line took him seven seconds *longer* than it had earlier. All those mental gymnastics had not improved, and had possibly enfeebled, his "faculty" of memory—that is, assuming such a thing even existed, which was starting to appear doubtful.

Young Thorndike was enthralled. Philosophers had long buttressed the sagging notion of a faculty of memory with wordy arguments and flights of introspection, and it had stood imperturbable for generations. Then along had come a lone academic who, with a single swing of a well-aimed experiment, brought the whole thing crashing down. When Thorndike enrolled at Harvard to continue his studies in English and French, he made sure to take a psychology course with James, and within a year he'd set his sights on a doctorate in psychology.

Funnily enough, by the time Thorndike began pursuing his PhD, James was in the midst of a shift away from experimental psychology and toward a highly successful second career as a philosopher. Nevertheless, when Thorndike needed a place for his experimental chickens to roost—Harvard told him he couldn't experiment on human children, so he'd begun working with young chicks—it was James who provided the nest, permitting the fowl to live and undergo study in the basement of his family's home. Thorndike later wrote that he hoped the resulting "nuisance to Mrs. James" was "somewhat mitigated by the entertainment to the youngest children."

It was the start of several formative years of research around which Thorndike would build his most enduring theories of mind in general, and learning in particular. At the time, psychology was still mostly a qualitative pursuit, although there were a handful of notable, quantitative exceptions (including Jameses' memory study and the work of the German psychologist Hermann Ebbinghaus, another self-experimentalist, who catalogued how long it took him to forget long lists of nonsense syllables). Thorndike, however, was a zealous believer in the power of numbers—even over such unruly subjects as newly hatched chicks. In the Jameses' basement, Thorndike built mazes out of textbooks and sent his downy charges into them. The chicks with maze-running experience, he soon observed, would complete the course faster than their naïve colleagues.

Before completing his degree, Thorndike decamped (with his two "most educated" chicks, as he described them) from Harvard to Columbia, the leader in the nascent field of experimental psychology. In Manhattan, he set up shop in the stifling attic of Schermerhorn Hall to continue his animal experiments. He soon graduated from chicks to stray alley cats (and sometimes dogs and monkeys), and from mazes to "puzzle boxes" of his own design, made of wooden slats. Inside them, Thorndike locked his hungry study animals until they managed to open the box's door by such means as pulling on a loop of rope, stepping on a pedal, or performing several of these actions in a specific sequence. (Thorndike's son later recalled that these boxes "would have shamed Rube Goldberg.")

"You'd like to see the kittens open thumb latches, and buttons on doors, and pull strings, and beg for food like dogs and other such stunts," Thorndike wrote to Bess Moulton, the woman he'd eventually marry, "me in the meantime eating apples and smoking cigarettes."

Thorndike quantified the behavior of his would-be escapees, timing their progress while counting the behavioral actions that led up to each successful attempt. In a cat's first run-in with a puzzle box, escape usually occurred only after protracted trial and error, and was always a happy accident. In later attempts, however, as with the chicks in their maze of books, experience paid off. Up to a point, the more successful escapes a cat had under its collar, the quicker it escaped. Learning, it was clear, was taking place.

But how exactly had this knowledge made its way into the cats' heads? In 1898, at the astonishing age of twenty-three, Thorndike laid out a theory to explain his observations. By all rights, his report should have been a small thing—merely a precursor to the five-hundred-plus publications, including fifty books, he would go on to produce. And yet, as he wrote to Moulton, "I've got some theories which knock the old authorities into a grease spot."

Most egregiously to Thorndike's mind, the "old authorities" in the study of animal behavior had imputed powers of humanlike reasoning to animals. He found this ludicrous; his escapees' performance had improved only gradually, evidence that sudden epiphanies played no part in their learning process. In his 1898 dissertation, *Animal Intelligence* (later revamped in an extended 1911 version), he gave an alternate explanation: Animals only ever solved problems by means of trial and error, and committed their solutions to memory only by repetition, or the "wearing smooth of a path in the brain," as he put it. Decades later, neurological research would partially vindicate this idea, showing that the points of synaptic connection between neurons involved in memory storage do indeed grow stronger with repetitive activity—about as close as you could come, physiologically speaking, to a brain pathway "worn smooth" with use.

Eventually, Thorndike codified his observations into a set of laws governing learning. The most important of these, which he dubbed the Law of Effect, held that for an animal engaging in trial-and-error exploration, any actions that registered as having a "satisfying effect" become more frequent, while those leading to a "discomfiting effect" become less so.

The implications for human learning were staggering. *Homo sapiens*, he decided, though capable of flights of reasoning beyond the ken of his caged cats, still relied on essentially the same processes as his animal subjects when it came to encoding memories. "These simple, semi-mechanical phenomena," he wrote, "are the fundamentals of human learning also." Assuming other factors were aligned in favor of learning (a ready and willing mind, content-to-be-learned chopped up into digestible chunks, and a schedule of repetitive practice), then the assumptions behind the Law of Effect remained "the main, and perhaps the only, facts needed to explain" human learning.

THE BIRTH OF BEHAVIORISM

The Law of Effect cast two shadows deep into the twentieth century. One fell over institutions of mass education, where its influence lingers to this day. The other crept across the scientific field of experimental psychology, where it helped inspire a movement that reigned for decades. In a 1913 address at Columbia, the Johns Hopkins psychologist John B. Watson introduced the term *behaviorism* to describe the research doctrine that Thorndike and his Russian contemporary, Ivan Pavlov, had already mostly staked out. True behaviorists, in Watson's extreme definition, would seek to explain the mind purely in terms of stimuli (events acting on an animal) and responses (the animal's observable behaviors), ignoring or even denying the influence of intermediating mental processes. According to this theory of learning—somehow even stricter and more stripped-down than Thorndike's—the only thing that could reinforce a pairing between stimulus and response was the frequency with which they co-occurred. In an attempt to provide a behaviorist explanation for human phobias, Watson and his assistant, Rosalie Rayner, conditioned a nine-month-old child known to history as "Little Albert" to fear fluffy white lab rats (and by extension, other fluffy things) by making a loud noise every time one came near. Ultimately, it wasn't these experiments that got Watson fired, but rather his relationship with the twenty-one-year-old Rayner, which turned romantic, led to Watson's divorce, and caused Johns Hopkins to show him the door. He spent the rest of his career in advertising, and is credited with coining the term "coffee break" for Maxwell House Coffee.

Behaviorism didn't end with Watson, however. B. F. Skinner, the inventor of the aforementioned teaching machine, soon took up the standard, revivifying Thorndike's animal research in the form of his Skinner Box, in which animals performed various actions to earn a food reward. His feats with pigeons alone are worthy of volumes:

He taught them to play Ping-Pong, for instance, by training them to peck a rolling Ping-Pong ball off the far edge of a small table, and then pitting them against one another. Skinner's pigeon-training efforts reached their acme of usefulness when, in the early days of World War II, he developed a pigeon-powered guidance system for American glide bombs. Trained to peck a photograph of a target to be bombed, the birds were packed into a bomb's nose cone, where their heads were tethered in such a way that their pecking movements would steer the bomb toward the real target, on the ground. The scheme actually worked in trial runs, but was obviated by the development of radar and never put into action. Decades later, Skinner bemoaned the premature end of his pigeon program. It hadn't all been for naught, however: "We had been forced to consider the mass education of pigeons," he said, and the lessons he had drawn, like Thorndike's, applied across species lines. "There is a genetic connection," Skinner said, "between teaching machines and Project Pigeon."

He began work on his teaching machines after sitting in on his youngest daughter's fourth-grade arithmetic class. When one of his pigeons did something right or wrong, it received immediate feedback. In his daughter's classroom, however, children filled out worksheets in class only to receive feedback the following day, when their teacher handed the graded sheets back to them. Skinner's research suggested that this was too late, and so he began work on a machine* capable of providing instant feedback, which enjoyed nearly a decade of reasonably widespread use before it became a cautionary tale, relegated to university storerooms.

In a sense, however, Skinner had been right about one thing: The behaviorist revolution that Thorndike helped kick off hadn't fully penetrated the workings of the classroom itself, or the inviolable relationship between teacher and student, or students' innermost

* Strictly speaking, a handful of teaching machines and related patents predate Skinner's—most notably, the machine developed by Sidney Pressey, a Jazz Age psychologist influenced by Watson and Thorndike. His invention, an intelligence-testing machine that also happened to correct wrong responses, only ever sold 127 units, however, perhaps because it was expensive and the Great Depression was unfolding.

thoughts. Rather, like a pigeon-containing missile, Thorndike's legacy, in the form of standardized educational structures, surrounded these things on all sides.

THORNDIKE GOES TO SCHOOL

In 1900, more than a decade before Watson coined the word *behaviorism* and several years before Skinner's birth, Thorndike accepted a professorship at Columbia University's Teachers College, where he turned his attention away from comparative psychology and toward human education. His Law of Effect provided the conceptual underpinnings for what would become a decades-long, nationwide (and ultimately international) push for standardization across primary and secondary schools, colleges, and universities. Throughout his long, exceptionally productive career, Thorndike would remain an enthusiastic participant in this process, lending intellectual heft to like-minded figures in the realms of science, education, and government, known collectively to education historians as the *administrative progressives*. As they made their stamp on mass education, the Law of Effect would both inform their decisions and, just as frequently, provide their maneuvers with ex post facto justification.

The administrative progressives were part of a far larger social trend, known as the Progressive movement, in which institutional actors sought to impose order on a society that seemed to be going off the rails. The apparent chaos manifested in a variety of ways, including urban migration, recessions, the new economic concept of "unemployment," and the concomitant rise of both the modern corporation and organized labor—forces that came into sometimes-bloody conflict. Of particular interest to educators was the disappearance of once-reliable escalators to middle-class prosperity, caused by both the advent of corporations and the national market economy. Though you might be middle class, the working world would make no such guarantees for your children.

It was, in retrospect, inevitable that educators would get involved in trying to fix things; even today, education presents a tempting, unitary point of influence over a whole host of social problems, with potential downstream effects touching everything from childhood poverty to economic malaise to racial inequity, not to mention existential concerns like falling behind rival countries technologically. The administrative progressives harbored no doubt about the seriousness of the problems they were facing, nor that they were the right people to do something about it.

In theory, that "something" could have resembled the holistic approach to education Dewey had tried out in his Laboratory School. Practically speaking, however, it's hard to imagine the gear teeth of the larger Progressive movement meshing with any approach other than Thorndike's, with its insistence on making learning measurable, countable, and mass-replicable. As the education historian Stephen Tomlinson has written, Thorndike's views provided the administrative progressives with "the tools necessary to atomise, sequence and monitor every aspect of schooling."

Dewey's influence would live on, mainly in teacher-education programs, which would provide a long-standing back channel into classrooms. But front offices—from individual schools on up to entire departments of education—would fall in love, and in line, with Thorndike. His Law of Effect would justify the standardization of curricula, the standardization of school administration, and, most disastrously, the standardization of students.

—

Perhaps the first casualty of Thorndike's theory of learning was the traditional, classical humanist curriculum. The old idea that skills bled across categories—that if you honed your "faculty" of memory on Greek poetry, say, you could later bring it to bear in your career as an accountant—made no sense in a model where information was stored in minute, stand-alone associations. And if "faculties" like memory, reason, perception, and judgment weren't improvable (Thorndike experimentally attacked the faculty of "judgment" with particular zeal), the logical move for school administrators was to cater directly to students' future endeavors. The four curricular pillars of mathematics, science, English, and history survived this shift unbroken but not unbent. Old standbys such as physics and trigonometry gave way to "general science" and "general mathematics" courses meant to apply straightforwardly to "the home, the farm, the nearby factory, the municipal and water plants," as a 1920 National

Education Association report put it. History, reformulated as "social studies," took on responsibility for churning out good citizens. Classical languages fell by the wayside in favor of such inventions as "home economics" and "physical education," as well as vocational training in fields like industrial drawing and woodworking.

By the 1910s and 1920s, a growing set of administrative superstructures were enforcing these sorts of new norms—to Thorndike, a strong step in the right direction. He contemplated a world where teachers would cede entire control of the educational machine to administrators and curriculum creators (Thorndike, who made a fortune producing textbooks and dictionaries, being a standout example). He even anticipated a far-flung future where a hypothetical algorithm could take the yoke out of teachers' hands entirely, anticipating "the effect of every possible stimulus and the cause of every possible response in every possible human being," he wrote in 1906. Whether it was administrators or scientists or machines delivering edicts from on high, his plans left teachers with a limited role, lower in the hierarchy. As Ellen Condliffe Lagemann has pointed out, gender was a major component of this teacher-diminishing worldview: administrators were often men and teachers women.

Such ideas couldn't have resonated more perfectly with the tenor of the time. A national vogue for efficiency and quantification was in full swing, and bespectacled, waste-hunting men bearing stopwatches and notebooks had become a common sight in offices and factories. In schools, too, pressure mounted to be efficient with time and taxpayer money. Larger school districts gobbled up their neighbors while professionally trained administrators, many hailing from Columbia's Teachers College, gently coaxed the reins of day-to-day operations from the hands of local school boards.

To this new administrative class, familiar tools like "credits" and "credit hours" became indispensable: common coins of educational attainment not only proffered by students moving within and between institutions, but also invoked on education's supply side in matters ranging from scheduling to teacher pensions to facilities construction. The standardization of educational time was essential to the administrative progressives' larger project. Unfortunately, as I explore in the chapters ahead, it came at a real cost to the biological imperatives of learning in the brain.

As Thorndike and his allies gathered influence, however, any natural obstacles to learning that their system might impose became less of a concern. In fact, learning itself was losing its totemic significance at school in favor of another function: sorting worthy students from unworthy. "Grammar school, high school, and college all eliminate certain sorts of minds," Thorndike wrote in 1901, a winnowing process that he embraced as a feature, not a drawback, of a successful education system. To him, the main challenge facing schools was not to improve intelligence, but to separate the apt from the inept. This was the only way to give the subset of students "who most deserve education beyond a common school course," he wrote in 1906, "such a training as will make them contribute most to the true happiness of the world."

Later, in 1924, he drove this idea home with a highly influential study, nominally an investigation into whether newfangled school subjects or their time-honored predecessors more effectively raised students' intelligence. The answer, Thorndike announced, was "none of the above"; what mattered more than courses studied was the mental equipment of the students doing the studying. "Those who have the most to begin with gain the most," he declared, a finding that lent urgency to the administrative progressives' commitment to sorting students. Looking back on the study today, as the education historian Herbert M. Kliebard has written, "there may be some question as to whether Thorndike was warranted in drawing such sweeping conclusions." In a very real sense, however, Thorndike's conclusions were foregone, having been formed decades earlier.

In the early 1900s, he and other educational psychologists had endorsed the use of subject-specific standardized tests to compare students, which public schools readily adopted. (Thorndike didn't invent the idea of standardized tests, but he did personally come up with a number of them, including scales for the comparison of children's skills in spelling, handwriting, history, English comprehension, and drawing.) These were soon overtaken, however, by the one mental evaluation to rule them all: the IQ test, which Thorndike embraced wholeheartedly. Although he and the IQ test's primary mover, Stanford's Lewis Terman, disagreed slightly about what intelligence was, exactly, they agreed that for all practical purposes it was determined at birth, and IQ did a decent job of approximating it. They joined forces with other like-minded educationalists and psychologists and pushed to make IQ testing the scholastic norm.

As IQ's influence grew in schools, debates about its efficacy became heated. Dewey, for one, sided with the opposition, arguing against any "procedure which under the title of science sinks the individual in a numerical class . . . and thereby does whatever education can do to perpetuate the present order." Despite such admonitions, IQ testing became the standard means of student-sorting in Jazz Age schools and a major preoccupation of the psychology research community at large. Terman became the president of the American Psychological Association in 1922. In 1925, the U.S. Bureau of Education determined that 64 percent of elementary schools, 56 percent of junior high schools, and 40 percent of high schools were using intelligence tests to classify their students into tracks aimed at either college or blue-collar work. And, as the education historian and analyst Diane Ravitch has written, the decision to plop a student onto this track or that "became a self-fulfilling prophecy, since only those in the college track took the courses that would prepare them for college."

—

To what end, all this quantification, this categorization, this winnowing on a mass scale? The effects, as I've mentioned, weren't all harmful; even the crudest standardized measures of student aptitude will result in fairer outcomes than admissions based solely on who-knows-which-alumnus. And yet, at the same time, whom did these supposedly meritocratic innovations weed out?

Let's start with what might by now be obvious. The entire intelligence-testing project of the early twentieth century was racist and classist—not just in execution, but also in motivation. The impulse to weigh and rank students, part of the general Progressive vogue for quantifying everything under the sun, was also tied more directly to another movement having a moment in the early twentieth century: the abhorrent, so-called science of eugenics.

In fact, both threads—behavioral statistics and eugenics—trace back to a single person, the British polymath Francis Galton, who can fairly be called the father of both. (He even coined the term *eugenics.*) Administrative progressives, especially Terman and Thorndike, venerated Galton; looking back in a brief 1936 autobiography,

Thorndike declared that, together with those of William James, the writings of "Galton have influenced me most." In addition to endorsing Galton's eugenics notions outright, both Terman and Thorndike also permitted a deeply unfortunate combination of racial bias and genetic determinism to influence their educational outlook. Terman, a particularly virulent eugenics advocate, was convinced that testing the intelligence of different races would reveal discrepancies so vast they could never be resolved by any educational intervention. And Thorndike, who belonged to New York's pro-eugenics Galton Society, endorsed a combination of segregationism and vocationalism: for instance, changing "certain schools for Negroes from a predominantly literary to a predominantly realistic and industrial curriculum." Even as late as 1940, when anthropologists had revealed that there was far more to human nature than what was written in our genes (while geneticists, wielding new chromosomal research findings, had demonstrated that there was far more going on *in* our genes than eugenicists understood), Thorndike still endorsed eugenics. "One sure service of the able and good is to beget and rear offspring," he wrote. "One sure service (about the only one) which the inferior and vicious can perform is to prevent their genes from survival."

Today, there is no shortage of research disputing the assumptions underlying Thorndike's and Terman's racial theorizing. In fact, as I've already mentioned, environmental factors have an enormous effect on performance on even modern IQ tests—sufficient, in the cases of underserved races in the United States, to more than eclipse any hypothesized group differences. What's more, marginalized populations *around the world*, as the anthropologist John Ogbu has written, tend to score ten to fifteen IQ points below local dominant groups. These differences disappear when both groups migrate to a new setting, the only plausible explanation for which boils down not to genetics but rather to the experience of marginalization. Or, as *Vox*'s Ezra Klein has put it concerning the United States in particular, the real issue isn't "that IQ isn't heritable, or even that it's impossible to imagine it differing among groups. It's that it's impossible to look at the cruel and insane experiment America has run on its black residents and say anything useful about genetic differences in intelligence."

—

Thorndike, the scientist who more than anyone brought the study of learning into the quantitative era, built a reputation as a dispassionate logician—especially when compared to the more anthropological and introspective Dewey. With all his ideological background details filled in, however, Thorndike begins to look less serenely rational and more like any other hot-blooded human, complete with an all-too-human tendency toward motivated reasoning and confirmation bias. He and his fellow administrative progressives permitted just-so stories about race and intelligence to interfere with their ability to abstract truths from the world around them. (Their willingness to entertain myths wasn't limited to intelligence, either. Thorndike used an index of class and race to draw an American map of purported "goodness.") You could be forgiven for laughing at this sort of scientific hubris—until you remembered the serious harm such hubris visited upon the world.

We're still living with the legacy of the Progressive Era's bunk science. To this very day, our systems of education are charged with a dual set of goals that sit in uneasy coexistence. On one hand, the primary goal of mass education is still to teach: to confer useful, activated knowledge on everyone. On the other hand, our secondary mission, inherited from the administrative progressives, tells us to standardize education's content, delivery, *and even its recipients*—which often runs counter to the ideal of learning for all.

This low-simmering state of dissonance runs hot wherever the machinery of standardization clashes against the reality of how learning *actually works*. As I've mentioned, the cognitive science of learning is a multilayered affair, comprising far more complexity than Thorndike's base-level theories account for. In the chapters ahead, we'll climb up through the layers of this cognitive high-rise, with each level expanding our scope. We'll begin at the microscopic scale—how does learning register in the very cells that compose our nervous system?—and ascend until we reach rarefied levels encompassing metacognition (how we think about thinking) and social dynamics. Each layer imposes new conditions on learning; addressing these conditions may lead us to a more inclusive, effective vision of education for educators and learners alike.

I hope it's clear by now that I hold Thorndike's legacy in mixed regard at best. And yet as we move forward, we must reckon with the uncomfortable fact that (at least in terms of methodology, if not ideology) our coming voyage will have undeniable Thorndikean aspects. True, we're adding layers to something he saw as two-dimensional, but still, we'll be seeking to understand learning by breaking it down into its nuts and bolts—an underlying impulse he would have recognized immediately. Thorndike's reductionism was an extreme form of an approach I'll refer to as *inside-out thinking:* attempting to explain complex systems as mechanistically as possible. Inside-out thinking offers a powerful way to abstract truths about the world, and an even more potent way to engineer theory back into practice. That is, assuming the model you develop is faithful to reality, which, in Thorndike's case, wasn't quite true.

Dewey's more holistic ethos, meanwhile, is an example of what I'll refer to as *outside-in thinking:* experimenting on a complex system *as a whole*, without cracking it open. This relatively cautious approach is less susceptible to ruinous error, but is also substantially harder to convert into educational practice at scale.

It's clear to me that any truly ambitious attempt to propagate learning of the caliber I've witnessed in 2.007 will require both inside-out and outside-in thinking. Together, these seemingly contradictory impulses may enable a more intentional approach to education, where we break with the past as necessary, but without jettisoning the secret sauce of the classroom altogether, as in Skinner's teaching machines. The next two chapters, dealing with the lower, more fine-grained levels of the cognitive high-rise, will necessarily reflect an inside-out ethos. Then, as we climb higher, we'll begin to stray into the realm of outside-in research. As we'll see, these two strands of scientific thought can lead to wildly different recommendations for educators, ideas we'll have to reconcile if we hope to offer a clear vision for how best to nurture learning.

For our first chance to challenge the status quo, however, we won't need to climb far at all. In the next chapter, I'll describe how, even at the fundamental level where Thorndike's theories still hold the most water, the biological reality of learning intrudes, and hints at a better way.

- II -

LAYER ONE: SLUG CELLS AND SCHOOL BELLS

"Space out your studies!"

"Don't cram for exams!"

"Start studying before *the night before!"*

I don't know if there is any truly universal experience in this vast, variegated world, but hearing the above injunctions must come close. How many of us, over the decades, have been told that it's a bad idea to limit our studies to the night before a test? That it's a hallmark of laziness, of poor time management, moral decrepitude? That, down the road, it leads to poor retention, diminished prospects, occasional dyspepsia, sweaty palms, epidemic unhappiness?

And yet, despite the direness of these warnings, many of us persist in making the wrong choice. I know I did, well into adulthood. So do many of my students, despite the fact that I'm now the one urging them to space out their studies. And so do students everywhere. In one study from 2013, half of the college-age students surveyed said they commenced studying only a day or two prior to their exams (and, bear in mind, these are the students who *admitted* it), despite the fact that the vast majority said they knew better.

Indeed, study sessions separated by days, even weeks, are dramatically more effective than studying all at once. Spaced learning, alternately referred to as "spaced repetition" and "spaced practice," is up there with vigorous daily exercise and flossing in the universally-acclaimed-yet-avoided department. If its benefits came in pill form, as they say about exercise, we'd all take it every day. The spacing effect is one of the longest-researched topics in cognitive psychology, and a wide variety of experiments, ranging back to the German psychologist Hermann Ebbinghaus's studies of his own recall in 1885, have shown it to work for all sorts of knowledge, in all sorts of students.

Want to learn Swahili? When researchers broke a vocabulary study session into two sessions separated by a one-day gap, students remembered 34 percent more Swahili words come test day, a week and a half later. Want to learn math? Four weeks after their lessons ceased, college students who practiced problems in spaced installments performed twice as well as those whose studies were lumped into one long session. Researchers have observed similar results among science students, too. Spacing has been shown to work in adults as old as seventy-six as well as in children, and even infants—and not just in academic realms of knowledge either, but also in unplanned-for, incidental learning, and even for motor skills, redounding to the benefit of everyone from expert pianists to novice golfers learning to putt. It also works in areas where bookish knowledge and motor skills intersect, such as the surgical operating theater. In one 2006 study, surgeons learning to interconnect rats' blood vessels who were taught in sessions spaced out over weeks performed significantly better than a control group given massed practice (that is, the opposite of spacing). In fact, only in the massed practice group did some surgeons make such a hash of their work that they had to give up on their tiny patients.

Wide-ranging though the above examples may seem, they account for a vanishingly small segment of the spacing effect's true sphere of influence, which runs much, much further afield than the human species. Our fellow mammals, too, seem to benefit from spacing, which might not come as a shock—they are, after all, our close relatives. Perhaps more surprisingly, however, relatively simple invertebrates also profit from a little space in their education. Fruit flies can be taught to fear certain odors, for instance, and this memory proves stickier if their training sessions are distributed. But still, fruit flies at least *have* a brain. Astoundingly, even animals without a brain or central nervous system can still benefit from spaced repetition. Take the sea slug *Aplysia:* a graceful, squishy genus, typically the size, shape, and color of a well-done pork chop. It has no central nervous system to speak of, and relies instead on a distributed net of neurons. Even these seemingly unsophisticated animals, when taught to react to different experimental stimuli, can remember what they've been taught for longer when that instruction is spaced out.

And *Aplysia* really is a simple species, neurologically speaking. The human brain contains 86 billion neurons. *Aplysia*, brainless—yet, one imagines, content, as it goes about its day grazing on marine algae—boasts just 18,000 neurons in its decentralized network. If you wanted to hand out a human brain neuron by neuron, you could give one to each resident of twelve consecutive Earths. By contrast, you could hand out *Aplysia*'s neurons as a participation prize at the Boston Marathon and still leave ten thousand finishers disappointed.

If having a neuron count just one five-millionth of ours doesn't stop *Aplysia* (and even simpler animals, for that matter) from benefiting from spaced learning, that means something. The fact that the benefits of spacing are so resoundingly conserved all the way out to the farthest reaches of the animal kingdom suggests that there is something essential about spacing. Long treated as a peripheral concern by educators—a good habit to promote, but not something to get worked up over—spacing in fact deserves more of a starring role in how we think about teaching and learning. As a cresting wave of research is now suggesting, it appears to be crucial to the very mechanisms that make memory possible.

CAJAL'S PROPHECY

The spacing effect wasn't always given short shrift in memory research. To Hermann Ebbinghaus, the bearded, bespectacled German researcher who helped found the field, it was the key to making memories stick. In a series of meticulous experiments he conducted on himself by memorizing long sequences of nonsense syllables, Ebbinghaus quantified forgetfulness over time, describing a "forgetting curve" in which most of the syllables faded quickly from memory, but a few lingered far longer. Repetition, he observed, could steel-plate memories for the long haul, especially if spaced out over time. This finding made a certain amount of intuitive sense: After all, he reasoned, "the schoolboy doesn't force himself to learn his vocabularies and rules altogether at night, but knows that he must impress them again in the morning." In 1890, upon reviewing

Ebbinghaus's research, William James proposed the existence of two separate categories of memory, which he called "primary" and "secondary." It was probably his student, the prolific E. L. Thorndike, who, in 1910, gave them the names most people would recognize today: *short-term* and *long-term memory*.

In the years that followed, psychological research into spacing didn't grind to a complete halt, but it didn't exactly set the eyes of leading memory researchers alight with passion either. To the behaviorists and their progenitors like Thorndike, spaced repetition seemed helpful but far from essential, providing a useful schedule over which to train rats, pigeons, and pianists, but little else. And so, as psychologists toiled to catalogue which stimuli led to which behavioral outcomes, spacing took on sideshow status.

The behaviorist-allied psychologists weren't the only scientists plumbing the depths of memory at the time, however. While Thorndike and his heirs did their best to understand learning from outside the skull, neurophysiologists were already poking around within. Soon enough, the physiological brain, so long a mystery wrapped in fascia, bone, and skin, began grudgingly to give up its secrets. By necessity, however, these early efforts were concerned with gaining the lay of the brain's terrain: what neurons looked like, how and when they seemed to convey information, how they connected to each other. With these sorts of questions still looming, the how and why of spacing would have to remain unanswered.

A separate question, just as intriguing but no less remote, was when the paths of memory research in psychology and physiology would meet. The behaviorists, to their credit, never doubted that at some far-off point, the corporeal structures underlying learned behavior would eventually come to light. Skinner, squinting at the hazy future, predicted, "What an organism does will eventually be seen to be due to what it is, at the moment it behaves, and the physiologist will someday give us all the details."

To midcentury neurophysiologists, meanwhile, the physical stuff of memory actually felt closer at hand—albeit wraithlike in its ability to avoid detection. Their research tradition began with Santiago Ramón y Cajal, the Nobel Prize–winning Spanish neuroanatomist whose painstaking drawings of branching, forking neurons remain in educational use to this day. In an 1894 lecture delivered in French to a group of London scientists, Cajal issued a prediction: Memory was made possible by the formation of novel connections—later named synapses—between neurons. "The cerebral cortex is similar to a garden filled with trees, the pyramidal cells,* which, thanks to an intelligent culture, can multiply their branches, sending their roots deeper and producing more and more varied and exquisite flowers and fruits," he told his rapt audience.

He was on the right track. In fact, if a neuroscientist today went back to 1894, sloshed Cajal on the back of the head with a blackjack, and delivered his lecture in his stead, that time traveler's message would differ in only a few key points. But Cajal's new-connection theory was untestable at the turn of the century, and in the meantime, other, competing explanations seemed just as likely. In the 1920s, for instance, based on the maze-navigating skills of rats missing chunks of their brains, the Harvard psychologist Karl Lashley supposed that memories were preserved by electric fields.

In 1949, Lashley's onetime student, Donald O. Hebb of McGill University, put forward a theory, more closely aligned with Cajal's, that turned out to be significantly warmer. The big problem, then as now, was one of complexity. In today's textbooks, neurons tend to be drawn simply, for ease of comprehension. As traditionally depicted, a neuron looks sort of like a tulip plant that's been plucked from the ground, root bulb and all, and laid end-to-end with similar plants in front and behind. At one end, emanating from our plant's bulb, a number of root-resembling structures called *dendrites* receive incoming signals from prior neurons. Once a signal is received by a dendrite, it travels the length of the neuron: past the bulb-shaped *cell body*, which contains the nucleus and other life-support systems; and along the narrow, lengthy stem, known as the *axon*. This transit occurs by way of an electrochemical mechanism known as an *action potential* (which is frequently analogized, to mix metaphors, to a line of falling dominoes). When the action potential reaches the flowery bits, known as *axonal projections*, these finally pass the signal along to the next neuron by dumping a payload of a chemical (a *neurotransmitter*) into the empty space (a *synapse*) separating the petals of our

* The principal cell of the brain's neocortex, first described in detail by Cajal.

neuron from the roots of the next. And so the signal continues ever onward: excelsior.

There actually are neurons in the body that look and act almost as simplistically as this textbook ideal: simple messengers that do little more than convey a signal from point A to point B. But in the brain, things are not drawn so clearly. A given cortical neuron is connected not at two points, fore and aft, but at on the order of *ten thousand synapses* (and sometimes as many as a hundred thousand), which connect it to many thousands of other cells. One of the toughest problems midcentury neuroscientists faced, therefore, was the question of when such a cell—not a tulip in an orderly row so much as a thorny shrub in the densest underbrush imaginable—receives an incoming signal, under what circumstances does it deign to pass said signal along? And, when it does, which of its many, *many* downstream neighbors receives the message?

Hebb suggested that every time one neuron causes another neuron to fire in a chain reaction, some self-reinforcing process must happen in those cells (and their shared synapse or synapses) to make that reaction increasingly likely to recur. The upshot was the existence of orderly, semi-discrete assemblies of brain cells, formed and maintained through the repeated usage of particular synapses. Or, as neuroscientists like to say today: "Neurons that fire together, wire together." The idea was revelatory.

More intriguing, however, was Hebb's corollary: Perhaps it was these web-like assemblages of cells that preserved memories in the warp and weft of their connections. As results piled up in support of Hebb's theory, neuroscientists began to circle. If the site of memory could be located physiologically, then it was just a question of how, by whom, and in what model organism: cat or rat, mouse or monkey.

ENTER *APLYSIA*

Or—though very few experts of Hebb's era would have predicted it—the sea slug *Aplysia,* known mainly for the simplicity of its nervous system. Behaviorism-inclined researchers approaching memory preferred complex animals: mammals and certain birds capable of impressive feats of problem solving and memory. Meanwhile, to the neuroscientists working in relatively simple animal models,* behavior was the hang-up. When you were peering through a stereoscope and dissecting out individual neurons, an animal's rationale for climbing a tree, grooming a mate, pushing a lever—that was all unassailably complex. As Eric Kandel, the neuroscientist who eventually earned a Nobel Prize for describing the connection between memory and the action of individual neurons, recounted over the phone in 2018: Most researchers "shied away from behavior. They thought it was too messy and too complicated."

Kandel was born in Vienna, Austria, but he and his family fled the Nazis in 1939, when he was nine years old, alighting in Brooklyn. By the mid-1960s, Kandel, who trained as a medical doctor before turning fully to research, had experienced neuroscience at the level of both entire brains and individual cells. After a maddening early experience at Columbia University studying the hallucinogenic effects of LSD, not yet a restricted drug, on the highly complex visual cortex of cats, Kandel yearned for a simpler study subject. He soon found himself drawn to the laboratory of a neighboring researcher who was probing the properties of crayfish neurons. There, he learned to manufacture glass electrodes, insert them into the crayfish's large axon—not quite the size of the squid giant axon, but substantial nonetheless—and connect them to a loudspeaker. The setup transformed silent neural activity into an audible cannonade. "I am not fond of the sound of gunshots," Kandel wrote in

* We owe much of our early knowledge of neurons and synapses to animals with simple nervous systems. Our first insights into action potentials, for instance, came from the giant axon of the foot-long squid *Doryteuthis pealeii*. Most squids have a giant axon running stem to stern, and the one belonging to the human-sized Humboldt squid is especially magnificent: the length of a tall person's arm, and, at one millimeter in diameter, a thousand times thicker than human axons. The longest axon in the animal kingdom is likely the hypothesized connection between the brain and fluke of the blue whale. For scale, imagine that axon in the Statue of Liberty, running from her head to her sandaled foot.

his memoir, *In Search of Memory*, "but I found the bang! bang! bang! of action potentials intoxicating."

A few years later, working again with cats, Kandel had the opportunity to use these electrode techniques in the hippocampus, the brain region that had recently been shown to be essential to long-term memory formation thanks to a series of landmark studies by McGill's Brenda Milner.* Since hippocampal neurons were required for long-term memory formation, Kandel's plan was essentially to poke around and see what made them so special. In short order, he and his research partner made a number of minor discoveries, but perhaps their most important insight came from the absence of a major breakthrough. The hippocampal neurons they were studying were special, it seemed, but not special enough to account for the hippocampus's unique role in memory. That suggested an intriguing possibility: The stuff of memory might not reside *within* cells so much as *between* them.

Against the advice of senior colleagues, Kandel swung his sights back to simple invertebrates. As he later wrote: "Few self-respecting neurophysiologists, I was told, would leave the study of learning in mammals to work on an invertebrate. Was I compromising my career?" But Kandel knew what he wanted: a study system that could do for memory what the squid giant axon had done for the action potential. Something, perhaps, like a certain sea slug Kandel had encountered in a pair of lectures by visiting scientists. "The advantage of *Aplysia* was, the nerve cells are gigantic. They are uniquely identifiable. And you could work out a simple behavior in terms of the neural circuit," he recounted. "If you know the neural circuit, you can see what happens to the neural circuit when the animal learns something."

The way he described it in retrospect, "see what happens to the neural circuit," sounded almost straightforward. At the time, however, it was tantamount to the holy grail of memory research, because if you could describe what happened to every cell in a circuit

* These concerned a man named Henry Molaison, known to the public only by the initials "H.M." until his death in 2008. More on Milner and Molaison in the next chapter.

where learning had occurred, you might just have a cellular explanation for memory.

—

Before he could join the French laboratory of Ladislav Tauc, one of the visiting *Aplysia* researchers, Kandel had to complete a two-year medical residency at Harvard Medical School, to which he had already committed. Residency was a more humane affair in the 1950s, however, and Kandel found it gave him time to read books, including B. F. Skinner's *The Behavior of Organisms*, which spelled out a number of Pavlov's and Thorndike's classic study protocols. "It occurred to me that the paradigms they described," Kandel later wrote, "could readily be adapted to experiments with an isolated *Aplysia* ganglion." With any luck, the type of inside-out learning model proffered by Thorndike and Skinner might even meet its long-dreamt-of biological mechanism.

After two years, Kandel and his young family made the trip to join Tauc in the seaside town of Arcachon. Finally, having stepped away for so long from the simplified neurological world of lobsters, crayfish, and squids, Kandel had returned—to reductionist, inside-out science; to the continent of his birth; to *les fruits de mer*.

SLUGGISH WORK

A little more about the magnificent creature that is *Aplysia*. Its enormous axons—the anatomical oddity that first caught Kandel's eye—are slightly wider than the Humboldt squid's giant axon, especially remarkable in an animal you can hold in your hand, as opposed to one you could hold only in a wet bear hug. When threatened, *Aplysia* can squirt ink with the best of them. They are also hermaphrodites, and form mating chains of as many as thirty individuals, with the slug at the back acting as a male, the one at the front as a female, and everyone in between pulling double duty. Sometimes, the locomotive meets the caboose, creating a pulsating, slimy ouroboros. *Aplysia* also exhibit an easily identifiable response to abject terror. In the wild, its spongy gill-and-siphon apparatus is exposed to the watery world most of the time, sticking out of its back like a backpack. Because the gill is both delicate and, to predators, delicious, whenever there's a hint of a threat nearby, *Aplysia* protects its gill by pulling it inward, into its body. Perhaps most important, this reflex can be trained—a fact Kandel would use to his advantage.

First, however, before working with the whole animal, Kandel focused in on *Aplysia*'s neurons. He removed an abdominal cluster of two thousand nerve cells, and, keeping them alive in aerated seawater, he sank electrodes into the biggest neuron he could find. He then stimulated smaller neurons converging onto the big one in an attempt, using just raw nerves and electrical signals, to replicate some of the most famous learning experiments from the history of the behaviorists.

There are three super-simple forms of learning, all of which were observed by the Russian proto-behaviorist Ivan Pavlov in his groundbreaking studies on dogs, which he conducted mainly by measuring their saliva production before and after various training regimes. The first, *habituation*, is a reaction to a constant neutral stimulus: for instance, when you move to a house near a rushing stream and gradually cease to register the sound of the water. The second, *sensitization*, is a response to an intermittent, usually noxious stimulus: when some unidentifiable device in your house emits an ear-splitting beep every fifteen minutes, causing you to jump higher each time. The third, known as *classical conditioning*, happens when a significant stimulus—something causing discomfort, fear, sexual arousal, etc.—becomes paired with a neutral stimulus. For an unsettling example, recall poor Little Albert, the toddler John Watson trained to fear fuzzy things.

Kandel achieved a habituation-like reaction first, by repeatedly triggering an incoming neuron to fire, which led to diminished responsiveness of the gigantic cell downstream. If an action potential is like a line of falling dominoes, the electrical value Kandel was measuring in the downstream cell, called the *synaptic potential*, is like the first domino's angle of tilt: push it beyond a certain threshold point and watch them all fall over. Using this metric, you can infer a synapse's strength; essentially, the more an upstream signal causes the downstream "domino" to "lean," the stronger your synapse.

Kandel discovered that under his simulated habituation, the synapse in question had weakened significantly, as expected.

Successful simulations of sensitization and classical conditioning soon followed. Kandel was now tantalizingly close to describing the cellular basis of real memories: reducing learned behaviors of the sort Pavlov had observed in living, breathing dogs down to the activity of *individual synapses.* This principle, however, could only be proven in living, behaving sea slugs, as opposed to the cell clusters Kandel had been working with—and in the meantime, his time in France was growing short. Upon returning to New York, Kandel assembled a crew dedicated to divining *Aplysia*'s remaining secrets. The team settled on the gill-withdrawal reflex as their primary study target, and Kandel used electrodes to painstakingly map out the neurons involved, which, he was pleased to discover, didn't vary from slug to slug. The real trick, however, was gathering readings of neural responses from live slugs, which Kandel and company eventually achieved by anesthetizing one, opening its neck, and pulling a still-connected abdominal ganglion out onto an operating stage. By 1969, the team had worked out how to measure the synaptic potential of the six motor neurons directly responsible for gill withdrawal. They began to train their slugs, electrodes at the ready.

And just like that, the dam broke. Sensitization and habituation, visible in the slug's gill-withdrawal *behaviors*, matched perfectly with the strengthening and weakening of the gill-withdrawal *synapses*. Classical conditioning proved tougher to crack, but the team eventually turned in similar results, showing that when they trained *Aplysia* to respond fearfully to a benign stimulus, it formed a chain of strengthened connections between sensory cells and motor neurons that had previously had nothing to do with one another.

In light of these stunning inside-out findings, long-simmering controversies began to fade. Lashley's electrical field theory of memory no longer made sense. Meanwhile, the far more esoteric argument over whether the human mind starts out as a blank slate, a position associated with the British philosopher John Locke, or comes partially preprogrammed, as the German rationalist Immanuel Kant advocated, became somewhat moot. If *Aplysia*'s neural setup and our own brains had anything in common, preprogrammed information did indeed come stamped into the anatomy of neural circuits, per Kant, while individual experiences, in accordance with Locke, determined how and when those neural circuits passed along signals. "The *potential* for many of an organism's behaviors is built into the brain," Kandel wrote. "However, a creature's environment and learning alter the effectiveness of the preexisting pathways."

REMEMBRANCE OF THINGS PAST

Perhaps the most important implication of Kandel's findings was that there was now a viable cellular mechanism for learning by association, the mechanism underlying E. L. Thorndike's theory of mind. In Kandel's experiments, however, the phenomena being associated were limited to simple stimuli and responses. How, you might reasonably wonder, can a system consisting of strengthened or weakened connections represent *ideas*? It's easy enough to imagine how someone might, say, sit on a cactus once or twice and develop an aversion to the sight of succulents. But how can such a system represent the idea of what a cactus *is*?

To explain, let's switch over from sight to smell, since visual signals undergo a great deal of processing on their way into the brain, while odor signals slip in with less modification. For a nose-wrinkling example, take the smell of an overripe banana. Let's suppose that you first experienced that unforgettable smell relatively recently—say, just one year ago. How, according to Kandel's model, could that smell have become represented in your memory?

When you first smelled that squishy, brown banana, a bouquet of volatile chemicals entered your nose and fit, lock-and-key-style, into some subset of your five or six million odor receptor cells, which come in some four hundred varieties. Those receptors that happened to match up with the volatile banana chemicals each fired off a series of action potentials that proceeded straight upwards, through perforations in the thin sheet of bone separating your nasal cavity from your brain, into the brain region sitting just on top of it, known as the olfactory bulb.

There, those incoming action potentials could have either triggered a second round of action potentials, leading further into the brain, or not.* Let's assume the former: The thousands of signals entering this first relay converged on a much smaller set of cell clusters, which reduced the cacophony of incoming signals to a relatively organized symphony before passing them along.

Here's where things got interesting. Eventually, the incoming odor signal reached a group of interconnected neurons that somehow penetrated into your conscious awareness when they fired. We don't know *how* activated cell assemblies trip into conscious awareness. (In fact, cognitive scientists disagree about whether we're even close to answering that question.) However, we do know that when groups of cells in certain areas of the brain light up with activity, sensations, ideas, and yes, memories, tend to fill our thoughts. In the case of our banana, when that assembly lit up for the first time, you experienced a new sensation, with some preprogrammed reactions mixed in. The sweet component of the smell, for instance, might have made you salivate.

If you were a simple animal like an *Aplysia*, the show would be almost over: You would act (eat the banana, run away from the banana), possibly develop some aversion or attraction toward future banana smells, and that would be that.

But you are most likely not a simple animal. In fact, you possess a brain that can be fairly considered the most complex stand-alone object in the known universe. And so when the banana smell came flooding in and started lighting up neurons, you didn't just act on it. You also created an internal representation of that smell in your head for future reference. Which meant that in addition to the nervous action going on at the ground floor of your brain, there was *another* assembly of cells lighting up somewhere upstairs.

In this model of memory (owing perhaps to a hypothesized ability of neurons known as *template matching*), such assemblies of cells will fire *only* if the right combination of incoming signals reaches it. Those signals can come from the senses—that is, the next time you

* Ever notice how you can quickly get used to a smell in a room, to the point where you can't even say whether it's still present? That's habituation, likely taking place in part at that first juncture in the olfactory bulb, which means that the synapse connecting the nose to the brain has become temporarily weak. For certain smells, this can come as a relief!

encounter a banana—or they come in from the side, via *associated* memory assemblies. After just a single banana encounter, you'll have associated a number of memories with the smell, such as the room you were in when you smelled it, who was with you, and the visual image of the fruit. You can also add additional associations after the fact. Importantly, a number of these associated memories will light up the next time you encounter a banana, and also the next time you simply *think* of a banana. Over time, the assemblies representing the smell of a banana, the sight of a banana, the word *banana*—these all become interconnected, essential components of what a banana means to you.

In this model of memory, then, the modulation of synapse strength of the sort Kandel observed in *Aplysia* permits an associative theory of mind recognizably similar to the one Thorndike postulated at the start of the twentieth century. Indeed, what made Kandel's approach so special was not his technical skill in the neuroscience laboratory so much as his interest in the sorts of simple learning patterns that had inspired Thorndike. In the 1960s and 1970s, as Kandel later explained over the phone from his lab at Columbia, he was far from the only neuroscientist able to place electrodes inside neurons. However, he said, "I was one of the few people in the world who was able to put an electrode into a neuron, who was interested in behavior, *and* wanted to analyze it in a cellular level."

"The reductionist approach that Thorndike, Watson, and Skinner led—that was very good," Kandel said. "But it doesn't give you *mechanism*. They simplify the behavior, and they show the behavior is altered. What you can do with biology, with reductionism, is get the mechanisms, and therefore go deeper into the problem." When asked whether the behaviorists *over*simplified, Kandel replied in the negative—their instinct to find inside-out explanations for learning wasn't wrong; they were just limited by the tools of early-twentieth-century psychology. "They just didn't go far enough," he said.

SPACING STRIKES BACK

In the list of mysterious phenomena that midcentury psychologists like Skinner hoped physiologists like Kandel would someday clear up, the spacing effect never ranked near the top. As the physiological work progressed, however, it soon became clear that there would be no explaining the mechanism of synaptic plasticity—changes in synaptic strength—without also accounting for the role played by time.

Of particular note was a finding Kandel's team made in 1971. If they stimulated *Aplysia*'s siphon forty times in a row, they discovered, it would lead to a one-day-long habituation of the gill-withdrawal reflex. But if they spaced out the protocol over the course of four days, then the habituation lasted weeks. It was already well known, going back to Ebbinghaus, that spacing out one's information intake had profound implications for the stickiness of the resulting memory, but no one had ever put a finger on where this attribute of learning resided biologically. Kandel and company's findings suggested its roots were very deep, indeed: conserved across species, and perhaps even isolable, like memory storage itself, down to the activity of individual synapses. No longer a sideshow of little relevance to more pressing neuroscience questions, spacing appeared to be at the very heart of memory formation.

But how, exactly, did synaptic strengthening work? A few clues were known already. In 1963, a team led by the married couple of Josefa and Louis Flexner had showed that mice, given a drug that interfered with their cells' ability to synthesize new proteins, were able to form new short-lasting, but not long-lasting, memories. This result hinted at two, maybe more, separate cellular mechanisms underlying memory: one that neurons could accomplish using just the molecular tools readily at hand, and one that required the neuron to fabricate new tools, in the form of proteins, from scratch.

Kandel's team replicated these findings at the level of individual *Aplysia* neurons and then began delving into the molecular mechanisms involved. The mechanism for short-term memory storage yielded first. Intuitively enough, it hinged on the amount of neurotransmitter dumped into a memory synapse by its upstream neuron: more neurotransmitter, temporarily strong synapse, temporarily strong memory.

Now the hunt was on to find the "switch" neurons flipped to make ephemeral memories more enduring. In 1973, a hint arrived from Norway, where the researchers Timothy Bliss and Terje Lømo had blasted a neuron in a rabbit's hippocampus with a high-speed train of electrical stimuli: one hundred jolts per second. Stunningly, the resulting boost in synaptic strength stuck around not for minutes, but for *days*. Onto this durable form of synaptic strengthening they bestowed the name *long-term potentiation*, or LTP. Like a well-preserved memory, the name stuck, despite the fact that, as became increasingly clear in the 1990s, LTP was really an umbrella concept, comprising multiple causal mechanisms and stages.

Today, LTP isn't the only candidate for how the human brain stores its lasting memories, but it's the clear frontrunner. The end stage of the process is fully reliant on the synthesis of new protein, just like long-term memory in lab mice, and now, thanks to advances in microscopy, we have a decent picture of what at least some of that protein is up to. LTP, it turns out, sends neurons through startling structural, anatomical changes. The local synaptic sites on the downstream neuron, known as spines, can increase in size markedly, adding neurotransmitter receptors along the way, which makes for a stronger synapse. Even wilder, whole new spines can also form, sometimes doubling up on the original upstream synaptic site, and sometimes forming *brand new synapses*, the effect of which may be to reinforce the original neural pathway multiple times over. Learning, it seems, doesn't just change your mind; it changes the literal structure of your brain.

The highly contrived, highly electrified conditions of experiments like Bliss and Lømo's don't occur in the wild, however, which raises the question of whether LTP really is responsible for memory, or if it's more of a laboratory artifact. The indirect evidence in LTP's favor includes its temporal milestones; like long-term memories, LTP is quick to form, long-lasting (persisting for a year in one study, and it may endure for longer), and its effects can be impaired in animals that have a diminished ability to learn (such as rats at the end of their natural lifespan). Meanwhile, when neuroscientists tailored their LTP induction protocols to mimic natural patterns of brain activity, it led to *stronger* synapses than standard methods—a hint that LTP can occur on its own, without electrode-toting scientists spurring it on.

LTP can also claim deeper, more mechanistic support. In one LTP-like mechanism Kandel's team discovered in *Aplysia*, a *third* neuron intervenes by delivering a trickle of the neurotransmitter serotonin to the upstream neuron of a memory synapse. This signal, passed forward by a chain of messenger molecules, snaps the memory synapse into a state of long-lasting robustness. Assuming something like this process also applies in our vastly more complex brains, it may help explain how moments of heightened emotion—waking up to the sight of a hungry monkey, for instance—can take on a flashbulb-like quality in retrospect, preserved in vivid detail. (I remember every tooth in that monkey's mouth.) In this model, your heightened emotional state acts as the trigger for the overriding signaling cascade, telling your memory neurons to retain incoming sensory information for the long haul.

This flashbulb effect is a useful, if sometimes trauma-preserving, feature of human memory. And emotion certainly has its place in education; an especially inspiring lecture, for instance, can stick with you for a lifetime. More often, however, we learners find ourselves ingesting facts and ideas without the mnemonic benefits of emotion. In these more workaday circumstances, learning is still, of course, possible—and we still likely have LTP to thank. In fact, perhaps the most important piece of evidence supporting an LTP-centric model of memory is the fact that, like long-term memory itself, LTP obeys the spacing effect. In the laboratory, spacing out LTP induction intensifies its physiological effects on both upstream and downstream sides of the synapse. Neuroscientists have proposed a few explanations for how this happens. In one persuasive hypothesis, only a subset of the molecular machinery that enables LTP is ready for action at any given time, a setup that rewards multiple, spaced-out encoding attempts, since each attempt gives you a fresh chance to recruit newly mature cellular components. According to this theory, the spacing effect may in fact be the *cause*, not the effect, of this cellular state of affairs. If information encountered repeatedly is more vital to survival than information encountered only once, then perhaps we animals have evolved a deep-seated filter to prioritize its storage.

There's more to learning—far more—than the encoding of memories at the synaptic level. In the chapters ahead, I'll talk about learning strategies arising at different levels in the cognitive high-rise: this one due to the structure of the brain, that one due to psychological motivating factors, and so on. What makes the spacing effect so special, however, is its depth. From the surface level of cognitive science, where students and psychologists alike can readily observe it, the spacing effect plunges down, down, down—to the very stuff of memory itself.

THE LONG WEIGHT

Sometimes, when I ponder my own relationship with learning, and how spaced repetition figured into it, I think about something called "the long weight."

After I graduated from university, having passed my remedial control theory course by the skin of my teeth (and following a short but pleasant stint at the University of Hawaii), I was lucky enough to land a job with Schlumberger, a company that builds and provides products for oil companies.

A few months later, I found myself strapped into a helicopter that had just departed Aberdeen, Scotland, clad in an insulated dry suit the company insisted I wear, ostensibly to give me a fighting chance in the frigid waters below in the event of a crash. The chances of something like that happening were remote, and yet I had to recognize that suddenly, for the first time, the stakes in my life had become tangible. My years devoted to vague preparation, to practicing for the real thing, were over. I pictured my helicopter silhouetted against the sun, like the famous image from *Apocalypse Now*. I was coming in to *do work*. My period of preparation was complete.

And what a period it had been. Once I'd signed on with Schlumberger, they'd whisked me off to their training facility in Edinburgh, where I'd spent the better part of two months preparing for my job in conditions that can only be described as deliberately annoying. An oil platform, as my fellow trainees and I were told, is a miniature, floating city designed for one purpose: to stab a glorified drinking straw through three hundred feet of moving water and a mile of earth and pull out a pressurized, explosive substance. Problems in such a system never stop arising, and our job would be to fix them—and on the problem's schedule, not our own. Which all sounded perfectly acceptable, until I discovered that our onshore training schedule would be every bit as unpredictable. To prepare us, Schlumberger had built what was essentially a mock drilling rig on dry land, complete with an enormous separator: a device with a tank the size of a smallish submarine that is used to vertically separate (hence the name) sand, water, emulsion, oil, foam, and gas. It relies on a complex sensor system that dangles down inside the tank. When everything is working smoothly, it lets you selectively pour oil from one spout, water from another. However, I can assure you, everything that can go wrong in a separator does indeed go wrong.

In our intensive training schedule, we'd spend a long day learning how some complex, computerized system worked, and then be released for a few hours into the black January night to blow off some steam pressure in the legendary pubs of Edinburgh. We'd come home, pile into our bunks, and try to sleep it off when an alarm would sound, the lights would come on, and our equivalent of a drill sergeant would march in to inform us that the separator was broken and we needed to fix it pronto—no time for pleasantries, no time for toiletries, now, now, now!

This sort of thing turned out to be typical. Almost daily, some system or other would get "broken"—actually sabotaged—which sometimes meshed nicely with what we'd been learning that day, and sometimes functioned as a refresher for something we'd learned weeks ago. Sometimes the "malfunction" would happen when the sun was up, but there must have been something about the wee hours, because more often than not, that was when wrenches appeared in the works. How suspicious!

Somewhere along the line, things began to click. Concepts that I'd vainly struggled to internalize at university began to simply make sense. A major part of this shift was due to seeing those concepts taken out of the world of textbook diagrams and put to work in the conduits, pipes, and valves before me, whose purposes I now understood. (I'll return to the importance of context for learning in the chapters ahead.)

But just as important was the matter of timing. Before, when I'd crammed for my exams, it wasn't because I was stupid or lazy. I did it because it *worked*—at least in the short term. Indeed, in most psychological studies of the spacing effect, cramming the night before an assessment is just as effective as spacing, and sometimes it can be even *more* effective on test day. But take that same test again after a week or a month, and the spaced-out strategy wins out virtually every time.

I'd understood that, in a general sense, at university. But I'd had limited time and attention, and once a final exam was over, there was never an immediate need to revisit the material. So I didn't. And in the long term, my retention suffered.

Only later would I begin to consider how curious it was that such unavoidable educational fixtures as final exams seemed almost custom-built to promote harmful study strategies. This discrepancy was, in fact, evidence of a collision point between two distinct functions of school: the promotion of learning and the standardization of education. My university's reliance on infrequent, high-stakes exams—as opposed to more frequent, cognitively friendly, lower-stakes assignments and quizzes—represented a victory of school's standardizing, winnowing function over the demands of students' learning brains. In fact, the entire overarching, regularized structure of university education, organized into semester-long courses full of information that, upon completion, students often never encounter again, runs contrary to the learning brain's need for spaced repetition. And while E. L. Thorndike can't claim personal responsibility for every standardized feature of contemporary education—the university term or semester, for instance, predates his influence—the biological importance of spaced repetition nevertheless adds an ironic twist to his legacy. School programs ostensibly set up to identify those students best able to form useful memory associations, per Thorndike's model of the mind, appear to step on the very cellular mechanisms that undergird associative memory.

My training in Edinburgh, by contrast, was like nothing I'd ever done, in that its unadulterated purpose was not winnowing, but rather the promotion of deep, contextualized learning. Knowledge that, in years past, I would have forgotten soon after I'd committed it to memory now stuck around as I re-accessed it. (Sometimes, like when I was roused from bed, I re-accessed it in a heightened emotional state, which—flashbulb effect!—only helped with retention.) Soon, I perceived the most wonderful thing happening. Have you ever noticed how, when you learn a new word, you begin to hear it everywhere you turn? This was happening to me, but with engineering principles. Before, I'd been building up stand-alone memory assemblies and letting their connections fade, sometimes over and over again. Now, I was building them up, reinforcing them, and involving them in *new* associations that, in turn, only added to the amount of regular exercise my assemblies received.

This was all taking place in early 1990. And so, unbeknownst to me, much of the research was still unfolding—often in the mind and laboratory of Eric Kandel—that would help explain why re-accessing and elaborating on memories strengthens them at a fundamental level, like how bones grow in the human body.

Today, we know that LTP induction (as observed in neuroscience labs) and memories preserved via spaced repetition practice (as observed in psych labs and research classrooms) are aligned, albeit imperfectly. For instance, both last longer the more spacing is used during training sessions, but only up to a point; there appears to be such a thing as too much space. In fact, there are still enormous holes to be filled in our understanding of the timing question. It's not known, for instance, how much spacing, and when, is ideal for LTP promotion in different types of neurons. Even psych research into the spacing effect has its limits. We know that if you want to remember something for a day, you should space your study sessions a day or less apart; if you need to know something for a month or more, you should space your studying out by a few weeks, maybe a month; and if you want to remember something for the truly long haul, you should revisit it on the order of months, maybe longer. But no psychologist can give you instructions more finely tuned than that.

That said, here's one good rule of thumb for spaced practice: If you're going over a topic and so little time has passed that it requires no mental effort to revisit it, then that doesn't count as spacing—that's closer to massed practice. Once you've done a little forgetting, however, and it takes some effort to relearn the nuances of a topic, that's a strong indication that you're on the right track. If you want to initiate the hardcore, anatomical changes of late-phase LTP, you need to convince your neurons that the information you've learned isn't something to be dismissed like every other inconsequential event in your day. One way to do so is by revisiting it when the memory starts to fade. Eventually, it will stick around for the long term.

But the need to build some space into one's study practice is no reason to stop and rest on your laurels. In fact, *interleaving*—the term for spacing out your studies by filling in the temporal gaps with different subjects you hope to conquer—doesn't just provide a time-effective way to space out multiple subjects at once. At the level of associative memory formation, it also allows you to make connections, where appropriate, between the different things you're learning, which can improve overall understanding and recall. (It also has other, higher-level benefits, which I'll explore in chapter 5.) In one clever 2010 study, for instance, two groups of fourth-graders practiced various types of geometry problems with the same amount of spacing in between each type, but in one group, the researchers interleaved the problems, while in the other, they delivered them in blocks, with spacing provided by filler activities. When tested afterward, the interleaved group performed twice as well.

This interleaved strategy, structured perhaps less formally, was precisely what Schlumberger subjected me to in Aberdeen. At first, I didn't like it. (This feeling is normal. In the study mentioned above, students in the interleaved group underperformed during their practice sessions; their advantage only manifested afterward.) Later, however, when my training was done, I felt extremely sure of myself. Barring a major disaster, it seemed there was very little that could go wrong on the oil rig that I wouldn't be able to suss out and fix. It wouldn't be overstating it to say that, when the helicopter touched down and I ducked out onto the salt-flecked helipad, I felt ready to kick some serious tail.

After a few extra days of onsite-specific training, I began my work. To start, I was told to shadow an older technician who would teach me the ropes. And indeed, there were ropes, or rather cables, everywhere: high-tension lines, pulleys, and weights, which sometimes had to be toted about by hand. At the end of my first day, my boss asked me to go get him a counterweight he needed. "Climb up that," he said, pointing to the highest spot on the platform, a metal structure accessible only via a long, slippery ladder, "and tell the guy up there that you need a long weight."

This request wasn't total nonsense: There were weights of all shapes and sizes on board, including, presumably, a long one. And so I did as I was told. I climbed the ladder in the freezing rain, hooking my harness to the metal framework in stages in case I took a spill, and then told the man at the top what I needed. "Sure," he said. "Stay put." And then he continued going about his business. I waited.

And waited.

It took me about twenty minutes to realize I'd actually been sent up there not for a long weight, but a long *wait*. I wish I'd had the good nature to laugh when I figured out the joke; I probably cursed. But I'd learned something: namely, that I had much more to learn. It was one lesson, emotion-tinged—and, in its way, spaced out—that will stick with me for the rest of my life.

THE CUTTING EDGE ADVANCES

In the years that followed, I began to look back on my oil-rig period with a certain nostalgia—and more than a tinge of disquiet. The training on offer in the petroleum industry is so effective because it has to be: The business is technical, competitive, and dangerous, and so its companies have no choice but to invest in their employees to a degree rarely matched in other industries. That's a nice perk for the individuals involved, but then again, you shouldn't have to work in petroleum in order to get top-notch, workplace-ready technical training. As I gained more perspective (and as the research on climate change continued to roll in), I found myself wondering whether the cognitively user-friendly aspects of my training experience could be freed—jailbroken, even—and made available to learners contemplating fields well beyond the world's fields of oil.

Today, owing in no small part to my North Sea experience, learning has become something I seek out for its own sake—not just for the knowledge gained, but also for the act of obtaining it. I find it's a decent personal ethos; more importantly, it's also the driving impetus behind science's continuing march into the unknown.

And the stuff of memory still certainly falls into the "unknown" category. Take, for instance, template matching: the notion that a given cell assembly will fire only after receiving a specific signal. The principle works really well in deep-learning computer algorithms, but we have no idea how, or even whether, it works in the brain. In fact, the synaptic-strength explanation of memory itself, though well supported, is still just the frontrunner in an ongoing race, and its victory is by no means guaranteed. As judicious observers have pointed out, most of its supporting evidence, though plentiful, is of the "strong-but-circumstantial" variety.

Very recently, a group of scientists here at MIT found a way to dig deeper. The story begins in the lab of Susumu Tonegawa, yet another Nobel laureate, who helped found MIT's Picower Institute for Learning and Memory. By 2012, there had been a long-standing push under way among the world's neuroscience labs to identify the specific neurons involved in a specific memory in a vertebrate brain. That year, a postdoctoral associate in Tonegawa's lab named Xu Liu, teaming up with a graduate student named Steve Ramirez, devised a way to do exactly that, in the brains of laboratory mice. They took a gene called *c-fos*, which becomes activated in neurons that have recently fired action potentials, and packaged in a few genes right next to it on its chromosome, so that whenever the cell expressed *c-fos*, the new genes would get expressed as well. (Imagine a vending machine that gave you a bonus pack of chewing gum every time you punched in the code for a Snickers bar, and you get the idea.) One of the new genes, borrowed from a bioluminescent jellyfish, coded for a protein that glows bright green in the dark. When the team shocked the foot of their genetically enhanced mouse, causing it to form a fear memory, its recently fired memory neurons began churning out this green fluorescent protein. Later, laid out on a microscope slide, the cells lit up like the Emerald City at dusk. It was the first-ever photograph of a memory.

But that was only half the story. That same year, a young doctoral student named Dheeraj Roy joined Tonegawa's lab. To explain the second half of the study, Roy led the way up to his lab space: a windowless world on the seventh floor of the Picower building where everything revolved around mice. The elevator was kept at a sweltering temperature—"to transport mice," he explained. "They get cold." The lab itself sat behind a security antechamber, where all comers were required to don a lab coat, nitrile gloves, and hair net for the sake, again, not of humans but of lab mice, which are often raised from birth in sterile settings and must be protected from the world outside.

The lab space, the size of a walk-in closet—real estate at MIT is often at a premium—was mercifully cooler than the elevator and equipped with four polycarbonate cages, each with metal grilles for floors and a pair of fiber-optic cables dangling portentously from their ceilings. Roy turned off the overhead lights and flipped a number of switches nearby. Circles of violent blue light appeared beneath each cable, on the floor of the cages.

The second half of Liu's study, Roy explained, relied on this blue light. In addition to green fluorescent protein, Liu and Ramirez had included a gene for a light-sensitive protein called channelrhodopsin, borrowed from a species of single-celled algae that uses it to find the sun. When a neuron, tricked into expressing this protein, is exposed to the right sort of light—specifically, the blue light issued from the fiber-optic cables—it goes wild, firing off action potentials like it's getting paid by the jolt. Since its invention in 2004, this technique, known as optogenetics, has brought unprecedented levels of precision to virtually every thread of neuroscience research requiring the external activation of neurons. In the team's case, it gave them the chance to selectively stimulate *only* those neurons involved in a brand-new memory. When the mouse received its shock and developed a fear memory of the cage where the shock took place, the neurons involved produced channelrhodopsin. Then, after putting the mouse in a different cage where there was nothing for it to fear, the team plugged the optical cables into a port in the mouse's skull and flooded its brain with light. The channelrhodopsin, triggered by the light, caused the fear-memory cells to fire. And the mouse froze in place, transfixed by the induced memory of its earlier shock.

Here the story takes a somber turn. Professor Liu tragically passed away in 2015, not long after accepting a faculty position at Northwestern University. He was just thirty-seven years old.

Depending where the research leads, Liu's legacy may nevertheless turn out to be enormous. In a study published in 2015, Tonegawa's team added a wrinkle to Liu and Ramirez's protocol. After training their genetically modified mouse to fear electric shocks, they gave it a drug that wiped out all built-up synaptic strength in its brain, thereby theoretically eliminating its memories. Soon enough, the amnesic mouse happily sauntered around the shock cage where it would normally have hunkered down in fear.

Their plan was merely to use this technique to add to the pile of evidence supporting the synaptic-strength theory of memory, and so they expected nothing much to happen when they switched the brain-lights on. They would stimulate the group of cells that had once contained the mouse's fear memory, but which was now, with its interconnections wiped, essentially just a haphazard smattering of cells. Even with those cells activated, they expected the mouse's happy behavior to continue, unchanged.

Which was why, when they flipped on the blue light and the mouse froze, so did the researchers. The implications were groundbreaking and immediately controversial. Memories, the team argued, must be encoded not only by the strength of synaptic connections, but also, somehow, by their connectivity *pattern*.

"The 2015 paper's conclusion was completely unexpected for everyone in our lab," Roy said. The team, now led by Roy, followed up with a pair of papers in 2016 and 2017 elaborating on their results, and the team's explanations began to look more and more plausible.

Not everyone agreed with them. Wayne Sossin, a neuroscientist at McGill University, acknowledged that the mouse's freezing behavior did indeed mean that a memory was being activated, but raised questions about the MIT team's conclusions. Tonegawa "talks about everything stored in the 'connectivity,'" Sossin said over the phone. But the difference between synaptic connectivity and synaptic strength? "To me, the two words are synonymous." Perhaps, as Roy and Tonegawa's team suggested, in the encoding of memories, new connections are forged between pairs of previously non-communicative neurons, and this new connection pattern, even in the absence of bulked-up synapses, holds the key to memory. Or, perhaps less excitingly, some small boost in synaptic strength was somehow left intact during the team's amnesia-inducing procedure, too weak to affect behavior under normal conditions, but strong enough to reveal itself under the inexorable exhortations of optogenetics. "I don't understand the mechanism that he's imagining for connectivity that's not a change in synaptic strength," Sossin said. That said, different mechanisms for memory, he acknowledged, could theoretically coexist. "It's not mutually exclusive. And I think one of the things that I like to emphasize is that not all synapses are the same. Not all connections use the same rules."

Memory at the cellular level continues to present a frontier for new discovery. But, rising above points of disagreement, one constant remains strong: the importance of spacing and repetition for lasting memories. In the Tonegawa team's new model, spacing only aids in the maintenance of new synaptic connections and LTP, both of which still appear to be necessary for natural memory recall. Sossin, meanwhile, has gone so far as to hypothesize a unified model of sorts that hinges on spaced repetition, with initial learning events causing new synapses to develop within memory circuits, and then spaced learning events stabilizing those synapses and using them to encode memories for the long term.

For too long, we've treated spacing as an educational add-on: an optional strategy that students can choose to ignore. But the spacing effect is not auxiliary to learning. Rather, it appears to be fundamental, present even in the minutest neural connections in extremely simple animals, inseparable from the stuff of memory itself. The temporal rhythms of memory are baked in: not just within our brains, but in our very heritage as learning organisms. Which makes me wonder: Shouldn't the way we teach better reflect this biological reality? There are certainly ways to shoehorn spacing into the traditional academic calendar as it stands—low-stakes quizzes that continually call back to earlier material, for instance, and increasingly cumulative exams—but my experience in Edinburgh and on the North Sea suggests that brave new educational structures, less beholden to norms and tradition, could go significantly further in realizing the benefits of spacing.

And perhaps they might carry other, higher-order benefits as well. If unnecessary impediments to learning exist at even the fundamental level of associative memory, what will the higher levels of cognitive science reveal?

- III -

LAYER TWO: SYSTEMS WITHIN SYSTEMS

Enter MIT's Building 46 from the south, and you'll walk under a sign that reads "Picower Institute for Learning and Memory." But if you walk around the side of the building and enter from the north, the sign overhead will read something different: "McGovern Institute for Brain Research." How a single building can contain two separate institutes makes a little more sense when you realize the structure is, in fact, a bridge. It straddles the Grand Junction Railroad, an eight-mile stretch of single-track, standard-gauge railway that serves as the lone, tenuous link between Greater Boston's southern and northern commuter rail lines. To provide necessary clearance for the trains passing beneath, the two sides of the building meet starting only at floor three: a tiled, echoing plaza, above which soars an atrium that widens with each level, like an inverse ziggurat. The expanse admits daylight into all corners, permits McGovern and Picower scientists to wave at each other, and also happens to weigh precisely nothing—which is exactly what you want when you're building over a void. Staring up at the glass roof, it's easy to forget that substantial parts of the atrium's floor and walls hang in the air, draped between the Picower and McGovern towers. Although each institute is capable, quite literally, of standing alone, when they are connected a monolith emerges, singular in form and function. Building 46 exists for one purpose above all others: to close a gap.

Cognitive science is a collection of fields separated by gaps in scale. The Picower is devoted mainly to neuroscience conducted at a fine, granular level: in and among the brain's individual synapses and cells. To get from there to the scale where most McGovern scientists operate, you'd have to zoom out. Far out, in fact. McGovern scientists routinely create digital images of the brain composed of 3-D pixels, known as voxels. Each individual voxel can contain as many synapses as a nine-hole golf course contains blades of grass. Many of the brain's great mysteries, including unanswered questions about learning and memory, exist between these levels of resolution. Thanks to work done at the Picower level, we know, at least in theory, how synapses can encode and hold on to information. And thanks to work done by McGovern scientists and their colleagues operating at what's known as the "systems level"—that is, in chunks of brain that would be discernible to the naked eye—we know of regions involved in the formation and recall of memories. In rare cases, we even have a sense of where cell assemblies dedicated to hosting certain types of memories seem to reside. But everything in between those levels of insight—which specific cells encode which memories, how those specific cells are chosen, the dividing line between one memory and another associated memory—these are the sorts of questions that we don't yet have the tools or methods to easily answer.

In fact, all of cognitive science is like this: a vast unknown punctuated with occasional scientific outposts. To explain, let's tip Building 46 over onto its side. Picture Picower falling facedown onto the street and McGovern rising into the air, trailing dirt and debris and colonial pottery. Now that we have two levels of a vertical structure of cognitive science disciplines, let's add more levels, above and below. For a basement, beneath the now-ground-floor Picower, we might add a level devoted to genetics and genomics, since one's genome wields considerable influence over neural development. Genetics tools overlap in some places with the ground-level work done at the Picower, such as in optogenetic experiments. But we're still not even close to bridging the gap between knowing what a few, or even a great many, genes do in theory, and how in practice that adds up to even a single living neuron.

The next platform up, the Picower level, benefits, as we've seen, from the techniques of cellular biology and fundamental neuroscience. It's sometimes possible to use these methods to stretch *up* and probe the darkness separating Picower and McGovern—to figure out which specific neurons do what in the brain—but it's an extremely painstaking and expensive process. Take, for instance, the team of neuroscientists who, in 2015, accomplished something similar to what Eric Kandel achieved in *Aplysia*, but in the fruit fly brain, a far more complex system. They isolated a key synapse that grew measurably stronger during a specific instance of learning, an amazing discovery. And yet, the looming goal of characterizing a complete neural circuit that fully accounts for a fruit fly's learned behavioral change, from stimulus to new behavior, is still maddeningly far off—and that's despite the fact that the entire network of fruit fly neural connections has been mapped out. Scientists still have to go through that "wiring diagram" bit by bit to see which synapse does what.

"I have a hope that we can figure out all the circuits from the sensory input to the behavior output," said the study's lead author, Toshihide Hige, "while I'm still alive."

Hige, it's worth noting, was thirty-seven years old at the time.

Happily, however, scientists are working to close such gaps in two directions at once, not just stretching upward from the cellular Picower level, but also downward from McGovern's systems level. Arguably, the marquee tool at McGovern is the fMRI scanner: the same doughnut-shaped, powerfully magnetic diagnostic tool you might find yourself occupying at the hospital, adapted to detect brain regions that are flush with oxygen-rich blood, a sign of local activity. Using this technology, scientists can pinpoint currently-in-use brain regions down to roughly the cubic millimeter. Cellular neuroscientists would find such brushstrokes absurdly broad. Considering that our heads contain more than a million cubic millimeters of brain tissue, however, for scientists studying the whole brain or large regions thereof, the relative precision and wide coverage offered by fMRI scanning can feel nothing short of miraculous.

The stack of cognitive disciplines continues upward, past McGovern. Imaging tools give way to a variety of psychological methods, and then, above that, sociological, anthropological, even economic inquiry. Knowledge gaps persist between higher levels, too: Cognitive psychologists, for instance, dig downward, stripping away confounding variables in order to peer into the workings of thought, while systems-level brain researchers work upward to try to see which surface phenomena, visible to psychologists, have deep roots, attributable to neural anatomy and function. The spacing effect is especially interesting because of its sheer depth, with roots plunging through regions visible and dark all the way down to the sub-synaptic level. Other comparable factors exist, however, and despite their shallower roots, they remain every bit as capable of making or breaking your chances of learning something.

Over a hundred years ago, E. L. Thorndike suggested that a person's knowledge consists of individual memories associated with one another in an ever-expanding web. It was almost as though he'd peered into a crystal ball and foreseen the state of affairs residing today at the Picower level: cell assemblies consisting of, and interconnected by, beefed-up or enfeebled synapses. What he couldn't anticipate, however, were all the processes originating *above* the synaptic level that would prove just as critical to when and how learning occurs—and, consequently, which students are found worthy or wanting in the educational winnower.

Today, we know far more about those intermediate levels, the obstacles to learning that can appear there, and what to do about them. In my mission to help more people develop a lifelong relationship with learning—expanding, along the way, our notion of who is worthy of educational investment—two systems-level research threads hold particular promise. One has to do with the physical architecture of memory storage in the brain. The other concerns how fundamental motivating drives, such as curiosity, intersect with those stored memories. Both threads loop multiple times through the McGovern Brain Institute at MIT. Let's catch hold of them there and see where they lead.

SWISS ARMY BRAIN

The tale of systems-level neuroscience in the past few decades, and its import for our understanding of learning, involves thousands of researchers. I'll focus on two of them, both at McGovern, who have elevated the idea of a highly compartmentalized brain—a model that lends itself well to an inside-out, mechanistic approach to improving learning.

They also both happen to have interesting heads.

That's one thing you might notice when talking with John Gabrieli: the levelness of his gaze. He keeps his head perfectly still, eyes fixed on yours. Perhaps it's a habit he picked up in the brain-scanning stocks of the fMRI machine in McGovern's basement, running preliminary trials on himself. It's as though there were something carefully organized up there, known only to him, that he doesn't want to throw into disarray.

Gabrieli is perhaps best known for his pioneering use of imaging technologies, which he's aimed at specific neurological disorders as well as at more fundamental questions, such as the mechanisms underlying memory and learning. These latter explorations have fleshed out our understanding considerably—and complicated matters commensurately.

There's more to the associations responsible for memory than I let on in the prior chapter. Remember that banana odor? In the interest of simplicity, I gave you the impression that the brain's association cortices (the poorly defined regions where representations of the banana's smell, appearance, name-as-spoken, and name-as-written reside) amount to an undifferentiated mass of synapses where memories can be stored essentially indiscriminately. Picture a slovenly private detective's office, littered with files flung so haphazardly that only their owner knows where to find them, and you get the idea. That impression isn't wrong, exactly, so much as outdated: It was the leading theory of memory organization in the first half of the twentieth century. Although Karl Lashley came up with the theory in the 1920s, he nevertheless spent a good chunk of the 1940s trying to disprove it, systematically damaging lab animals' brains in an effort to locate the site of memory traces. It didn't work. "I sometimes feel, in reviewing the evidence of the localization of the memory trace, that . . . learning is just not possible," he wrote in 1950. "Nevertheless, in spite of such evidence against it, learning sometimes does occur."

The first real indication that the brain relies on a more specialized filing system arrived in the person of Henry Molaison, who was known to the public only by the initials "H.M." until his death in 2008. In 1953, in a last-ditch attempt to correct his debilitating epilepsy, Molaison had matching chunks removed from either side of the center of his brain, a procedure that claimed most of his hippocampus and several nearby structures. The cure, which proved largely successful, came at a terrible price. He awoke into an eternal present, unable to form new long-term memories, a condition that persisted for the rest of his long life. This tragic outcome, instantly familiar to anyone who has seen the film *Memento*, transformed the science of memory. Initially, research spearheaded by McGill's Brenda Milner focused on Molaison's inability to form new long-term memories, which clued in neuroscientists to the importance of the hippocampus and surrounding structures for the process of long-term memory consolidation.* Since Molaison still retained memories from his youth, the hippocampus couldn't be the final resting place of long-term memory. Instead, researchers hypothesized, short-term memories had to somehow pass *through* the hippocampus, undergoing some sort of transformation along the way, before being filed away for good elsewhere.

By the end of the 1950s, it was becoming clear that there was at least some division of labor in the brain's memory systems, but the degree of systematization remained mysterious. Barring further evidence, it seemed that the organizational scheme might well turn out to be as simple as a pair of categories labeled "short term" and "long term."

Soon enough, however, those categories began to split and multiply like cells under a microscope. The focus of research into Molaison's brain began to slide from what he could *not* do to what he could *still* do, and it proved to be just as revealing. He was, it turned out, capable of improving his performance on certain motor tasks, such as tracing a five-pointed star on paper while granted only a disorienting, mirror-reversed view of his drawing hand. Although he

* Of the non-hippocampal regions damaged in Molaison's surgery, most, like the hippocampus itself, are found in the medial temporal lobe, a name that has nothing to do with time and everything to do with its proximity to your temples. His symptoms, however, were primarily due to his missing hippocampus.

could never recall practicing this odd task, his skill grew. Eventually, during one star-tracing session after much forgotten practice, Molaison exclaimed, "Huh, this was easier than I thought it would be."

The upshot of all this work was that long-term memory, once unitary, became bipartite. Consciously accessible memories of the sort Molaison reliably forgot—new facts, new words, new personal history—now fell within the category of *explicit memory*. Essentially everything Molaison still could retain, by contrast, became known as *implicit memory*, which came to include classical conditioning of the sort that Kandel had demonstrated in *Aplysia* as well as motor and perceptual skills: riding a bike, playing the piano, reading at full speed.

The idea of implicit memory wasn't new, exactly, just newly explicable. In the 1800s, a number of tales had emerged concerning memory in the face of amnesia. There was the British woman, for instance, who damaged her memory when she nearly drowned, and then learned the trade of dressmaking, despite the fact that every day on the job felt like her first. And then there was the forty-seven-year-old woman, amnesic as a result of alcohol abuse, who was pricked by the pin-wielding Swiss psychologist Édouard Claparède, subsequently forgot the attack, and nevertheless refused to shake his hand the next time she saw him.

Now explanations flooded in for these historical anecdotes as well as more commonplace observations, such as the fact that it's much easier to forget, say, the atomic weight of tungsten than it is to forget how to ride a bicycle. Once it was clear that these sorts of memories relied on different parts of the brain, their different degrees of stickiness began to make more sense. One paper from 1994, for instance, revealed that Molaison's star-tracing ability remained elevated even after a full year's break from practice. It was quite literally just like riding a bicycle.

The lead author of that paper was John Gabrieli. From the mid-eighties to the mid-nineties, Gabrieli and his doctoral advisor, the late, renowned MIT professor Suzanne Corkin, had essentially inhabited Molaison's implicit memory as they produced a lengthy list of what his brain could and couldn't remember. Their efforts received a welcome boost by a newly discovered form of implicit memory known as *priming*, which could be used to plumb subconscious memory associations.* In a typical priming study, researchers would measure the degree to which encountering a given cue—a spoken or written word, or, as the field evolved, other stimuli—made it easier, minutes or hours later, to come up with certain answers in response to experimental prompts. For instance, imagine I asked you to study a long list of terms, including the word *meter*, and then later, after you'd presumably forgotten that particular cue, I asked you to fill in the blanks in: "m e _ _ _" with your choice of either *meter* or *melon*. If you (or a statistically significant set of people like you) chose *meter*, that could be construed as evidence of priming. Somewhere in your mind, researchers hypothesized, a pathway remained oriented toward *meter*, and that lingering bias counted as an example of information storage in the brain—aka memory.

The biggest downside to these sorts of studies was that in even the most meticulously designed experiments, researchers could never really tell which results were due to priming and which were due merely to half-forgotten snippets of explicit memory. Amnesic patients like Molaison who had no explicit memory, however, blew this problem out of the water. When Molaison displayed priming, it had to be the real deal.

Gabrieli kept pushing and the categories of memory kept dividing. It turned out that Molaison was capable of feats of priming that people with different memory deficits couldn't pull off. Molaison could be primed to solve ambiguous connect-the-dots puzzles in a certain way, for instance, which didn't work in people with Alzheimer's disease. Gabrieli took this finding as evidence for some unknown yet qualitative difference between memories of words and those of images.

By the end of Gabrieli's PhD program, however, newer, more accurate tools were becoming available. In November 1991, a computer-generated illustration of an unfortunate-looking man's bald head graced the cover of the journal *Science*. He had a clean slice

* Later studies would put too much stock in priming by suggesting that it could be used to influence behavior outside the laboratory; these findings mostly proved impossible to replicate. To Gabrieli, however, exploiting priming in this way was never the goal; it was just a tool for sounding out the substructures of memory.

missing from his upper-rear skull, as though he'd leaned backward (forgive me) into an airplane propeller. A few regions of the exposed brain glowed orange. It was one of the first fMRI images ever taken of brain activity corresponding to human vision, and it served as the opening shot in a revolution. fMRI's precursor technology, positron emission tomography (PET), took what amounted to minutes-long-exposure photographs of milliseconds-long neural flickers, which was like trying to photograph a hummingbird's beating wings using a Civil War–era camera. An fMRI "exposure," by contrast, could be made on the order of a single second (and as a bonus, unlike PET, it required no injections of radioactive materials). As the technique's advantages grew apparent, brain scientists began clamoring for scan time at their local hospitals' fMRI machines.

Like many young brain scientists and neuroscience-oriented psychologists, Gabrieli, who was hired at Stanford in 1991, resolved to master this new technology, which promised freedom from the catch-as-catch-can nature of working with people who happened to have brain lesions. As fMRI researchers slid their test subjects into their scanners and fed them different stimuli, a picture of a highly organized brain began to form that stood in stark contrast to the cluttered-office model of old. There were perceptual brain regions, it soon became clear, that responded solely to specific categories of stimuli. The first, and perhaps most striking, of these was discovered by Gabrieli's eventual McGovern colleague Nancy Kanwisher.

—

In the 1980s, Kanwisher was a member of the same graduate school class as Gabrieli; today, they're both professors in that same department at MIT. And like Gabrieli, she too possesses a noggin that may interest passing observers, but for different reasons. For one thing, hidden beneath her hair, her scalp is covered in imperceptible tattoos.

In the early days of transcranial magnetic stimulation, an experimental technique that involves stimulating the brain with magnetic fields, Kanwisher found herself in need of exterior landmarks for her explorations, and so, pragmatically, she and a graduate student had dots of different colors, including a couple visible only under a black light, strategically tattooed. Tattoo parlors were illegal in Massachusetts at the time, so they had to travel to Providence, Rhode Island, where a beefy, ink-covered artist did the precision work. Kanwisher laughs about it now: "That was like twenty years ago," she said. "It kind of failed. It didn't work for shit, but it was amusing."

Before that, in the early 1990s, she was one of the hungry young scientists fighting for scan time in Boston. The pressure was on to come up with a big result as quickly as possible; she didn't have a major research grant, MRI usage was expensive, and a number of competitors were vying to snatch away her MRI access at Massachusetts General Hospital. She was searching for regions devoted to the visual recognition of shapes, but nothing was turning up in the scanner. To tighten her search, she dug into the literature concerning a region in the back of the right hemisphere that, when damaged, caused people to lose the ability to recognize faces.

"I had never worked on face perception because I considered it to be a special case, less important than the general case of object perception," she later recalled. "But I needed to stop messing around and discover something, so I cultivated an interest." She dove in: literally, into the middle of the fMRI machine, where she painstakingly tested her own brain's responses to faces and all sorts of other visual stimuli. A promising, glowing region appeared in the machine's mockups of the rear of her right hemisphere, and when it also showed up in other people's brains, she permitted herself to get excited. If the region, dubbed the *fusiform face area*, really was integral to the identification of faces, then possibly other, similar regions existed in the brain's perceptual systems—maybe even predominated. "This finding fit the broader idea that the mind is not a general purpose device," she later wrote, "but is instead composed of a set of distinct components, some of them highly specialized for solving a very specific problem."

Kanwisher only described the fusiform face area in 1997, long after fMRI's potential first became apparent. The reason it took so long had to do with the differences between individual brains. People's bodies vary as much on the inside as they do on the outside, and the brain is no exception—which is a problem when you want to locate the same tiny brain regions in different people. Systems have been created over the years to systematically describe brain locations, including a coordinate system that works sort of like latitude and longitude, but due to our natural variation, the same structure can show up at different coordinates in different people. What Kanwisher and others figured out (several teams stumbled onto this technique independently that same year) was how to locate brain regions *functionally*. Defined functionally, the location of, say, Boston wouldn't be 42.3° N, 71.1° W; it would be "the place where all the people live near the mouth of the Charles River." In the same sense, Kanwisher defined the fusiform face area not by its neural GPS coordinates but rather as the region in the fusiform gyrus* that responds significantly more vigorously to faces than to control stimuli.

Once you have your study subject's functional region mapped, then you can ask questions of it, like whether it responds to upside-down faces, or faces with no eyes. And so, painstakingly, you can start to home in on the filters that the brain's perceptual systems use to make sense of the world. Under this new form of scrutiny, highly specialized brain regions, once obscured by individual variation, began to emerge from the haze of imaging data like gorillas from the mist.

And that brings me to the other thing that's so noteworthy—and inspiring—about Kanwisher's head. In 2015, she was diagnosed with lymphoma, from which she has since fully recovered, but which necessitated a chemotherapy treatment that threatened to claim her hair. She decided to turn the loss into an opportunity to teach a video lesson on the brain's new functional anatomy. In a single, high-stakes take that has since been viewed more than 200,000 times on YouTube, Kanwisher gestures to a 3D model of her brain, rotating on a screen behind her, which has a number of her own personal specialized regions highlighted. "Where are each of these regions inside the head?" she asks the camera. "Well, it's kind of hard to tell with all the damn hair in the way." And with a decidedly un-Gabrieli-esque flip of the head, she gathers up a lock of hair into her fist and cuts straight across with scissors. Fast-forward a minute and she's seated on a swivel chair, completely bald, with her graduate student

* The ridges and canyons that give the outside of the brain its wrinkled appearance are known as gyri and sulci, respectively.

spinning her around, drawing standard anatomical regions on her scalp in black, and, in red, blue, purple, and green, Kanwisher's own specific functional regions—regions only uncovered in the wake of her description of the fusiform face area.

Today, there appears to exist a whole suite of these perceptual functional regions, many of which run along the side of the brain: one for music and another right next door for pitch; another for bodies; another for places. From an evolutionary perspective, most of these make intuitive sense: You can imagine how it might be advantageous to know, without having to think about it, whether a face belongs to someone you recognize or not, or whether a brandished object constitutes a threat. Because these sorts of visual filters would presumably have benefited our evolutionary forebears, there's currently a robust debate ongoing about the degree to which, say, the filters involved in facial recognition are the product of instinct or experience.

However, there is one functional perceptual region that can't possibly be innate, because it responds to stimuli that only came into existence five thousand years ago. Tucked in snugly next to the fusiform face area sits a tiny region known as the visual word form area—or, more colloquially, "the brain's letterbox." It is the part of the brain that scientists believe permits the near-instantaneous recognition of letters. Not only does such a surprisingly specific region exist, but it is shockingly invariant: barring serious medical problems, it's always found in the same place across individuals, cultures, languages.

"Why does that land in a systematic location?" wondered Kanwisher. "To me that doesn't make sense. Right? Like, given that there isn't a prior evolutionary history of people reading, how does it land always in the same place?"

In a dramatic departure from the messy-office model of information storage, a highly specific category of memory—the shapes of letters, and the identities of small groups of letters that signal the sound and meaning of written words—needs to be filed away in a very specific location. And if that doesn't happen properly, or if the connections servicing the letterbox and other crucial language-processing regions develop atypically, the ramifications can be significant.

The most common example of this is dyslexia: reading impairment at the level of individual words. As the 1990s wore on and the receipts continued to come in regarding the role of the letterbox and other regions involved in reading, the attention of systems-level researchers began to swivel in interest. Even the level head of John Gabrieli began to turn.

THE DYSLEXIC BRAIN

Gabrieli had spent the bulk of the 1990s running all things imaging-related at Stanford, where he focused on memory—especially areas where memory intersected with medical issues: Alzheimer's, amnesia, schizophrenia, Parkinson's, Huntington's. All the while, one non-medical area continued to hold his interest: reading, a holdover from his Molaison research. By the late 1990s and early 2000s, a number of baseline rules concerning the functional anatomy involved in reading had become clear.

Research in patients with brain damage, and in patients whose brains had been electrically stimulated during surgery, had long hinted at the existence of general provinces involved in the production of speech, the auditory recognition of speech, and visual word recognition. What these early findings—going all the way back to 1861 in one case—couldn't tell us, however, was how these parts worked together. By the 1980s and 1990s, brain imaging was changing the game, making it possible to visually trace some of the sorts of connections that Gabrieli had been exploring via priming. Sounding out the linkages in the auditory processing of words, for instance, might once have necessitated the tactic of asking amnesic patients to choose between homophones (primed with "taxi," would they later write "fare" or "fair"?). Now, however, an fMRI researcher could present study participants with either a written or a spoken noun, and then scan their brains for regions that became active when they read it silently or out loud, or responded to it with a related verb. Taken together, this sort of work suggested the beginnings of a complex flowchart for how written and spoken information shape-shifts around the brain.

The most important thing to know about this flowchart is that there are parallel routes involved. Let's try it out with a simple written word, like *peanut*. Presumably, when you read that word, a number of associations come to mind: the appearance and taste of the legume, baseball, Mr. Peanut, and so on.

How does your mind jump from the word on the page to the *idea* of a peanut? To start, the image of the written word enters your brain via your retinas, where it undergoes a series of basic visual processing steps. The recognition of the sorts of shapes used in written language selectively activates a pathway leading to the brain's letterbox region, which serves as a filter for incoming letters, identifying which are present and in which order. It efficiently differentiates among similar-looking-but-different letters (*C* and *G; i* and *j*) while bundling different-looking-but-the-same letters (*g* and *G; a* and *A*). The letterbox also doesn't discriminate between handwriting and print, or size of type. All told, the letterbox imports letter-shaped lines and turns them into letters-in-the-abstract. These are then organized into small groups, including *graphemes*, which correspond to the letters' sound-as-spoken (in "peanut" there are five graphemes: p-ea-n-u-t), as well as *morphemes:* the smallest unit of written language that carries meaning (in our case, "pea" and "nut"). These sorts of processing steps then inform further brain activity, which proceeds along two general routes. Both operate simultaneously, but one or the other picks up the bulk of the work, depending on what kind of word you're reading. If it's a familiar word like *peanut*, most of the action follows what's known as the "deep" reading route: where the letters on the page communicate more or less straightforwardly with the cell assemblies that correspond to the "meaning" of the word, which are believed to live in a distributed network known as the *semantic lexicon*.

The other, "surface" route, as opposed to the deep route, takes a less direct path to the semantic lexicon—but more on that in a minute. First, it's worth mentioning what we don't know about the role of memory in reading. For one thing, we don't know where any given entry in the semantic lexicon lives. The same is true for facial recognition: we don't know where our actual face representations reside. "I can close my eyes right now and imagine my mother's face. Boom, there it is," said Kanwisher. But "where is that memory stored? We have no freaking idea—and I find that scandalous! I mean, it is so embarrassing and scandalous to me that after 20 years of working on this, I have no idea. But I don't, and I think the field doesn't."

We do know, however, that each functional region lights up selectively for its preferred stimulus, and moreover, damage to such areas tends to result in specific forms of "blindness." A lesion in the fusiform face area can cause *prosopagnosia*, aka "face blindness," or the inability to recognize and differentiate between faces. And damage to the letterbox can cause *alexia*, aka "word blindness": the acquired inability to read. Even if the representations of faces and word meanings live somewhere unknown in the brain, the perceptual filters needed to process incoming facial and written information likely exist within the fuzzy borders of their respective functional regions. And since the filters for written language are certainly learned, not innate, it's likely that, in the brain's letterbox, we have the resting place of a very specific type of memory. The messy-office model of memory organization, then, falls away in favor of something closer to what Kanwisher calls a "Swiss Army knife" model: a brain filled with specific tools that perform specific functions. In the brain's letterbox, memory itself becomes a tool. Its nebulous, bushy associations become hammered flat, honed into a surgical blade that, no matter your native language, is always found in the same spot, and always serves the purpose of cleaving abstract information from shapes. In a very real sense, then, everything you've ever read and subsequently remembered was built using tools constructed from prior memories. When you stop to think about it, the whole memories-upon-memories structure can start to seem so precarious, so improbable, that you, too, might want to think twice before making undue head movements.

And indeed, your instinct wouldn't be wholly wrong, because as in any complex system, problems do crop up. The 5 to 12 percent of children who experience reading difficulty due to dyslexia can quickly find themselves with relatively sparse semantic lexicons relative to their age cohort, which makes it harder to read later in childhood, which in turn makes it harder to learn other things via reading. There are multiple causes of dyslexia (multiple dyslexias, really), but many cases appear to involve the atypical development of long, physical pathways in the brain: information superhighways connecting distant language-processing centers. Surprisingly enough, the reading route most affected in dyslexia doesn't appear to be the deep route running from the letterbox to the semantic lexicon, but rather the more circuitous surface route I mentioned earlier, which first leads to representations of the *sounds* of syllables and words, and only then proceeds to their meanings. Assuming you don't have dyslexia, this pathway leaps into action when you encounter a new word for the first time and sound it out—which, when you're first learning to read, happens frequently. Later, the auditory middleman can be cut, and reading can proceed more automatically.

In developmental dyslexia, however, things happen differently. There are a few hints as to why, visible way up at the psychological surface levels of cognitive science. Someone with dyslexia might, for instance, have trouble saying what the word *game* would sound like without the *g*, or struggle to sound out made-up words. With the tools available at the systems level of neuroscience, it becomes possible to see where, anatomically, these effects might arise.

In 2000, Gabrieli, as part of a team led by the Swedish cognitive neuroscientist Torkel Klingberg, aimed a new technique at this question. Diffusion tensor imaging, a cousin of fMRI, analyzes the predictably random movement of water molecules for non-random trends—such as those that occur when water molecules are confined to the tiny tubes that are neuronal axons. The technology's introduction in 1994 led to a series of breakthroughs in our understanding of how different sections of the cerebral cortex connect to each other. The cerebral cortex, which is the site of just about everything brain-related I've discussed in this chapter, consists mainly of a thin layer of neuronal cell bodies, known as gray matter, that blankets the whole surface of the brain. As Kanwisher puts it in her head-shaving video, the cortex is about the size and thickness of a large pizza, and the folds of the surface of the brain help the entire pizza fit inside the confines of your brainpan. Just beneath the cortical surface is a different tissue, known as white matter, which consists less of neural cell bodies and more of their long, tubular axons, which act like a sort of switchboard connecting different chunks of gray matter in all different directions. With diffusion tensor imaging, it became possible to map out bundles of white-matter axons, which, sure enough, turned out to connect brain regions that seemed to correspond to different reading-related tasks. In their 2000 paper, Gabrieli's team revealed something strange: A super-long-distance white-matter pathway, apparently involved in reading, appeared to be directionally disorganized—woolier, bushier than you'd expect—in dyslexic brains. A series of follow-up studies undertaken by Gabrieli and others ensued, and continued after Gabrieli joined MIT's McGovern Institute for Brain Research in 2005. As it currently stands, there are four major suspect white-matter pathways, two of which run essentially the full length of the brain, that appear to be less organized in dyslexia. One of these, known as the left arcuate fasciculus, is deeply involved in *phonological awareness*, or the ability to mentally manipulate the sounds of spoken language, a crucial part of the surface route in reading.

"It actually connects the posterior—the back part—of the reading network with the front, and that's quite interesting," explained Nadine Gaab, Gabrieli's former postdoctoral student and current frequent collaborator, a professor at Harvard Medical School and Boston Children's Hospital. Information flow between those areas is necessary for achieving reading fluency and reading comprehension, she said, and abnormalities in that flow could potentially account for many of dyslexia's manifestations.

PULLING UP THE ROOTS

In addition to disorganization in the brain's sinews, Gabrieli and his collaborators have raised another possible explanation for dyslexia. Both the symptoms and associated imaging results may be related to an underlying problem with synaptic plasticity: the ability, discussed in the previous chapter, of synapses to selectively strengthen and weaken. In this hypothesis, this seemingly general, systemic problem only affects reading because of all the things we do, reading is spectacularly, perhaps *uniquely*, demanding. "There is no other human behavior that approaches reading's demands for coordinating multimodal perceptual representations and cognitive processes," write Gabrieli and his coauthors. "In this way, a general neural dysfunction that is subtly detrimental to other behaviors may be substantially detrimental for learning to read."

Stepping back, it may seem strange that there are multiple conceivable points of failure affecting reading, and reading only. After all, it's not something we evolved to do. Most other things we've started doing recently in our species' history—cricket, video games, needlepoint—don't each have their own specific learning disorder, do they?

In the case of reading, there is so much that can conceivably go wrong because it's not a new application of classic tricks (swinging a cricket bat is not so different from swinging a wooden club), but rather a *hack* of our preexisting neural equipment. It goes so far beyond the design specifications of our brain's architecture that, if our heads came with a warranty, reading would probably render it void. We're already pushing the envelope of brain function so aggressively, it seems, that the loss or compromise of *any* of the biological machinery involved can easily imperil the whole enterprise.

To return to the brain's letterbox, a growing contingent of researchers has begun to suggest that its location is so consistent precisely because *it couldn't possibly exist anywhere else:* those letter-detecting perceptual filters must be situated where fibers carrying shape-related information from the eyes converge on connections leading to both the shallow and deep reading routes. In 2016, a mammoth team-up of scientists—featuring Gabrieli, Gaab, *and* Kanwisher—produced something close to a definitive explanation for the letterbox's mysteriously consistent location. By comparing connectivity patterns in the brains of pre- and post-literate children, only the latter of which contain a letterbox region (remember, the letterbox, like the fusiform face area, is defined *functionally*, so it only appears after you learn to read), the team was able to demonstrate that the presence of connections in the younger children presaged where the letterbox showed up a few years later. As they wrote in the paper, "the functional fate of a given cortical region" may be preordained by "its connectivity fingerprint."

Put another way, the hunk of long-ago-evolved, shape-recognizing cortex in which the brain's letterbox selectively squats is a highly specialized piece of neural machinery, built to translate visual shape information for the use of other regions near and far. So particular is this setup, in fact, that one school of thought even suggests that, far from our brains adapting to process written language, perhaps history has selected for the survival of written languages that meet the strict requirements of our brains. In this theory, the surprisingly short list of line shapes that form the backbone of virtually every written language—including those with characters—exists for the simple reason that those are the sorts of shapes that our visual processing system can parse quickly and automatically. "Reading itself," writes the French cognitive neuroscientist Stanislas Dehaene, "progressively evolved toward a form adapted to our brain circuits."

But don't mistake those circuits' specificity for total inflexibility. In the case of dyslexia, it is possible to mitigate its effects, especially if you catch it early enough. The trick, once a catch-22 but no longer, is to identify children at high risk of developing dyslexia—*before* they learn to read. Thanks to our growing understanding of the role played by speech and auditory processing in dyslexia, it's become possible to identify pre-literate children who might benefit from specialized interventions through such tests as asking whether two words rhyme, or by asking them to name a series of images in rapid succession. And now, in an effort to catch dyslexia at an even earlier age, Gaab has developed a computer app that turns such tests into games involving zoo animals. The app's main screen, a forest path, looks just like the parkside view outside her window at Boston Children's Hospital.

There is no doubt that the brain, especially the young brain, can be helped to overcome challenges posed by dyslexia. The mechanism for how these tactics work, however, remains mysterious. In the most straightforward explanation, practice simply strengthens the synapses of anatomical structures that would otherwise display reduced activation. "But just as often," said Gabrieli, people with dyslexia "also see the apparent growth of pathways, let's say, in the right hemisphere, not their typical reading hemisphere." This sort of development is evidence of a rerouting of information. "The thought is maybe you're not so much regularizing the left hemisphere reading pathway as you are promoting the development of an alternative pathway," he said. There are "some hints that some of the best outcomes in dyslexia are those who develop an alternative pathway, as opposed to modulating the typical pathway."

The recruitment of alternate pathways also occurs in dyslexic learners struggling to read on their own, but, in the absence of expert help, this unaided process is usually frustrating. A region in the left inferior frontal cortex associated with syntax and speech often goes into overdrive, which doesn't help—it is, as Dehaene writes, "a brave but often fruitless endeavor"—but it does testify to the effort exerted by many learners with dyslexia. The disorder, striking at the systems level of neuroscience, hinders even the most motivated learners, and it's been shown time and again to have nothing to do with general intelligence or aptitude for learning.

It is not alone, either. In fact, reading is not the only highly demanding learning task that requires the precise interplay of memory and perception. The same can be said for at least one other human endeavor: mathematics. Indeed, a consensus is coalescing around the existence of a sibling learning disorder of dyslexia, known as dyscalculia.

"It's a smaller area, a more recent area. In many ways, it's kind of like a fifteen-years-later version of dyslexia," said Gabrieli.

"Up to forty percent of kids with dyslexia also develop dyscalculia," said Gaab, "so there must be some overlapping early mechanisms. But we don't know anything about it."

How did dyscalculia remain hidden for so long? The answer may be as simple as the fact that you can get away with being "bad at math" in a way that is not true of reading. "If you can't do math, you pretty much can be very successful," Gaab said. "You can be a professor of English at Harvard, right, without doing any math. But if you can't read, you will have a really hard time."

As both dyslexia and dyscalculia become more comprehensible to neuroscientists, their existence raises important questions about whom our standardized education systems have historically permitted to advance, and whom they have winnowed. Turn-of-the-century ideas about intelligence and aptitude, simply put, were not built to deal with something like dyslexia, which becomes explicable only when you move up a level on our cognitive high-rise: when you stop worrying about *how* memory associations form and start grappling with *where* they form. As a result, dyslexia has interfered with individuals' educational trajectories ever since the modern rules of school were built on E. L. Thorndike's template. Only recently have we begun modifying these rules to be more forgiving, such as by granting students extra time on exams. For all we know, dyscalculia could be having a similar effect, undiagnosed and unacknowledged.

What else, lurking out there in the darkness surrounding systems-level neuroscience, might be concealing human learning potential from the systems that are supposed to detect it? What leviathans swim past the windows of McGovern at night, felt but never seen? Or, considering the situation more optimistically, what bounties exist out there, waiting to be reeled in, that can make learning more user-friendly, more effective, more inclusive?

READINESS TO LEARN

The outline of one of these figures flitted past Gabrieli's lab in 2011. That year, Julie Yoo, a postdoctoral associate, spearheaded an investigation of the parahippocampal place area, a functional region first identified by Kanwisher and the University of Pennsylvania neuropsychologist Russell Epstein that appears to do for places and scenery what the fusiform face area does for faces. Yoo and the rest of Gabrieli's team (including Gabrieli's spouse and frequent collaborator, the neuroscientist Susan Whitfield-Gabrieli) were hoping to find out whether activity patterns in the brain could be used to predict a state of "preparedness to learn." The open question had puzzled scientists going as far back as Thorndike, who, in his canonical "Law of Readiness," had observed that there were moments when an inexplicable switch seemed to flip in the brains of his test animals, and learning, normally "satisfying," became "annoying."

Identifying a ready-to-learn state by direct fMRI observation of the full brain was impossible—there was too much brain to watch, and the relevant patterns might well be tiny—but the small size and high specificity of functional areas presented a unique opportunity. The team narrowed their gaze to just the parahippocampal place area and monitored it for activity patterns, while their MRI-bound research participants attempted to memorize and recall pictures of scenery. The approach paid off: The team identified an activity signature that predicted how well study subjects would remember a given scene.

The study was simultaneously groundbreaking and of little

practical use. You could theoretically find a readiness state in the brain, but you'd have to be teaching students lying inside an MRI machine to take advantage of it. And anyway, the study's findings only applied to memory for places, not facts. More than anything, it was a proof of concept. "Can you measure, in the brain, states of brain function that are conducive for a specific kind of learning?" Gabrieli asked. "As modest as that goal sounds, that hadn't been done before."

Mission accomplished, Yoo moved on to another job, and Gabrieli swung his sights back to dyslexia and other disorders. When I became aware of the study, however, it stuck with me—not so much due to the idea of a brain state conducive to learning, but because its existence suggested the opposite must exist as well: a state of unreadiness. Although it was not, strictly speaking, a disorder, since it likely affected everyone, "unreadiness to learn" was still a threat. Like dyslexia, it could potentially throw education off track. It, too, seemed to have roots extending down to the systems level of cognitive science. And as with dyslexia, we didn't have to accept its effects.

—

As I've already discussed, my academic performance suffered in my final years at university, in part because of the timing of how the material was taught. There was more to it than that, however, which I now believe had to do with a particularly stubborn form of unreadiness to learn. My difficulties came to a head at one specific point in the semester. My professor—a leading light in his field; it wasn't like my university was skimping on instruction—was teaching my classmates and me about the relationship between fluid dynamics and the speed of sound.

At a surface level, supersonic flow isn't really that complicated. Basically, when you have a fluid moving through a tube and a partial blockage occurs—a pinch in a garden hose, a control valve in a pipe—it creates a choke point, through which the fluid can shoot at extremely high speeds. The lecture wandered off into the subject of shockwaves, which I got, but couldn't figure out why they would matter for any sort of engineering that didn't have to do with supersonic aircraft. My professor also made a vague reference to the "back-propagation of information," but I didn't understand how it fit in.

Between that lecture and the subsequent exam, every time I cracked open the relevant section of my textbook, my eyes rolled into the back of my head. I knew I should want to figure it out, but it just . . . wouldn't quite fit. The experience, I decided, was like tissue rejection. It was as though my mind couldn't find a way to make this new information matter in a useful way, and so classified it as a foreign object.

Fast-forward a year to my job on the North Sea. The rig I was working on wasn't the permanent kind, built into the seafloor, but rather a floating platform designed for exploration, typically by plunging a pipe into the seabed and measuring what comes up. Geologists can divine a lot about what's underground by the pressure of the initial stream of oil produced and how quickly it tapers off. The steadier the pressure, the more oil is likely down there, but measuring this value can be an uncertain business, since the changes you're looking for can be minuscule, and in the meantime you're standing on a sloshing watercraft filled with exceedingly complex equipment maintained by, in my case, profoundly inexperienced deckhands.

I was working on this system when I had an epiphany so great I nearly fell into the sea. When the oil flows to the platform, it passes through a closeable choke valve. The platform's geologists monitored the pressure of our exploratory well at a point upstream of this valve, but pressure fluctuations emanating from downstream of the valve had the potential to throw off their readings. To avoid this, my colleagues and I were told to "choke the flow," which would cause the flow to move faster than the speed at which sound travels in oil. That was when I finally got it. "Sound" is actually made out of pressure waves: when you speak, you're sending a signal of high- and low-pressure pulses through the air. A sound wave (or pressure wave—same thing) that moved upstream through the column of oil and reached the geologists' instruments could easily throw off their highly sensitive readings. If we kept the flow choked, however, and the outflow fast enough, a pressure wave could never advance past the choke point; it would be as futile as shouting a message at a retreating Concorde jet. Moving fluids do indeed carry information, I realized, assuming you're the sort of person who cares about pressure readings. And, just like my professor had said, that information can't back-propagate when a choke is tight enough. Suddenly, I was eager to retrace my steps and figure out all the other details on the subject that I'd tuned out at university. Because these events predated the World Wide Web, however, I had to content myself with badgering my more senior colleagues about it, which, to their credit, they put up with for far longer than strictly necessary.

Later on, when I'd calmed down, I wondered why I hadn't been able to understand what my professor was telling me the first time around. What had been the blockage? Had there been, dare I say, a choke point of sorts in my brain? And if so, what had opened the valve?

THE CURIOUS BRAIN

In 1994, the Carnegie Mellon psychologist George Loewenstein articulated a theory that hinted at an answer to questions like mine. When a brain detects a gap between the knowledge it contains and the knowledge it *might obtain*, he suggested, the not-altogether-unpleasant sensation known as curiosity can result.

One sort of curiosity manifests as long-term fascination with a given topic: Asterix comics, podcasts about grisly murders, sports news, that sort of thing. Loewenstein and the brain scientists who followed him wanted to know more about a different sort: a kind of fast-twitch curiosity. Let's say you just found out—oh, to pick an example at random—that there is a single known species of moth that would happily bite through your skin and drink your blood,* and its range is spreading. The impulse that is possibly sending you hurtling toward Google at this very moment is, in these scientists' reading, a drive state just like hunger, thirst, and the sex drive, as well as pathological states such as drug addiction. Entire subfields of psychology and neuroscience have developed around each of these drives, digging into what triggers them, when, and in whom; and

* *Calyptra thalictri*, found throughout Eurasia.

how they momentarily outweigh other desires. In the case of curiosity, where the substance desired is information, the vagaries multiply. What makes a given chunk of information feel highly desirable, and what renders others effectively inert—as useless to the curious mind as a handful of gravel to a hungry stomach?

By the mid-aughts, fMRI researchers were digging into the curiosity question and turning up clues. A small monetary reward for learning, a team including Gabrieli and Whitfield-Gabrieli determined, could spur activity in the hippocampus (the home of long-term memory consolidation) and cause information to linger for the long haul. They wondered if a similar, *intrinsic* anticipation of reward, such as that created by a state of curiosity, might have the same effect on memory storage. But the more they uncovered, the more challenges seemed to crop up. "Certainly, curiosity would be a form of readiness-to-learn," Gabrieli explained, but unfortunately, "the brain turns on in fifty different ways" upon encountering something it finds interesting, which makes it difficult to pinpoint what's happening mechanistically. Ultimately, curiosity's activity patterns proved so overwhelming that his group found itself looking for ways to *exclude* curiosity in their search for a single, stand-alone, readiness-to-learn brain signature. Conveniently, their study of remembered scenery made for "very bland materials," he said, which allowed them to identify a type of readiness to learn that appeared to stand independent of curiosity.

Other groups, meanwhile, took the opposite approach, attacking curiosity directly. A few years later, in 2014, a team out of the University of California, Davis, provided a tantalizing glimpse into what, precisely, curiosity does in the brain. While feeding trivia questions to their study participants, the researchers also showed them a constantly changing feed of photos of human faces. Afterward, the participants were more likely to remember a face if it had appeared during a moment of curiosity—curiosity, that is, for trivia information that had *nothing to do with faces*. When you're in a state of curiosity, it seems, the potential for long-term memory formation gets boosted universally, for all sorts of memories—a process that unfolds, in the words of the authors, "via dopaminergic facilitation of hippocampal LTP."

You read that right: LTP, or long-term potentiation, the synaptic star of sticky memory. As near as systems-level neuroscientists can figure, curiosity is like rocket fuel for LTP formation in the parts of the brain critical to long-term memory storage.

And thus, however tenuous the connection, research at the first two levels of our cognitive high-rise—cellular and systems—meet. The fact that curiosity is what brings them together is fitting. The difference between curiosity on an individual level and the collective, curiosity-fueled effort that is science, after all, is one of degree, not kind. It's all about closing information gaps.

The *how* behind the *why* of all this hinges on something I alluded to in the previous chapter: additional circuits of neurons that ride herd on those responsible for the grunt work of memory. Memory neurons communicate mainly through glutamate, which is by far the most common neurotransmitter in the nervous system. Higher, managerial circuits, meanwhile, wield other neurotransmitters that can have a variety of effects, as in the flashbulb-like memory storage Eric Kandel observed in *Aplysia*. In the case of human curiosity, it appears that a managerial group of neurons delivers the neurotransmitter dopamine to the hippocampus, causing memories there to become stickier.

What's so fascinating about these signals in the case of curiosity in particular, as opposed to hunger or thirst, is their point of origin. The hippocampus itself appears to determine whether incoming information is worthy of curiosity. Upon finding in the positive, it sends out excitatory signals to brain regions associated with the anticipation of reward; these regions then return the favor by sending dopamine signals to the hippocampus, telling it to turn new, incoming information into long-lasting memories. The hippocampus, in short, tells *itself* when to double down on information storage. The question of what triggers curiosity, then, may very well boil down to this: What sort of information, when acquired, causes the brain to hunger for more?

—

When I could simply not make myself feel curious about chokes in fluid dynamics, I was attempting to overrule some fairly fundamental drives in my brain. Both metaphorically and neurologically (given that curiosity resembles hunger), it was like entering a hot-dog-eating contest on a full stomach. I wasn't hungry for this knowledge, and I couldn't convince my brain to reward me for taking it in.

But on the oil platform, something shifted. With a bit of context, it seemed that I had crossed into a sort of Goldilocks zone, where I knew just enough to be able to file away more information—a state similar to what the influential Soviet psychologist Lev Vygotsky called the *zone of proximal development.* How, though, could one little injection of knowledge trigger a drive state in my brain?

The work of Jacqueline Gottlieb, a neuroscientist at Columbia University who specializes in curiosity, may contain an answer. In decades prior, she explained, the rules of thumb scientists believed to animate curiosity boiled down to either neophilia—the desire to explore anything new in your environment—or the so-called information gap hypothesis, which holds that curiosity ensues whenever the available information in your environment appears to be greater than what you already know. Both theories have drawbacks, however. The neophilia explanation ends where the dark, scary basements of the world begin; sometimes we're content to let the unknown remain unknown. The problem with the information-gap hypothesis, meanwhile, is that true information gaps are rare in the wild; more often, you don't know what you don't know. In a trivia contest, the missing chunk of information is clear: it's the answer to the question. There are certain similar situations with known unknowns, "like when you read a mystery novel, or when you watch a movie," Gottlieb said. "It's undeniable that you're riveted and you expect a certain informational flow." But far more often, whether you're a human in a classroom deciding what to do with your life or one of our primate relatives exploring a new environment, there are no such guarantees. "Curiosity is about learning, and it's about learning when you don't have a lot of constraints in the big wide world," Gottlieb said. In most situations, you're surrounded by far more data than you'd ever want to process, let alone remember, and so your brain must invoke curiosity only sparingly.

As a result, Gottlieb has suggested, a meta-strategy takes over—inclusive of, but not limited to, information gaps and neophilia. In what she calls the learning-progress theory, whenever you encounter a chunk of information that forces you to change or reframe your prior body of knowledge, curiosity is likely to follow. In the case of an information gap established by a trivia question, the information contained in the question suggests enough to draw you into a quest for more. And in the case of a more open system—the "big, wide world" Gottlieb alludes to—your curiosity flows only to the subset of available sensory information that augments or challenges what you already know.

A classroom, crucially, is an example of a big, wide world. A lecturer like me might think that he's the only source of information in the room, but even in the days before smartphones and laptops, that was never true. Other students, a bird outside the window, a fly in the light fixture—these things have always competed for students' attention. Even if a teacher tries to compel attention by force, the result is usually not curiosity but boredom: the bothersome feeling, to paraphrase the psychologist John D. Eastwood, that occurs when we can't engage with the information before us, even if we want to.

Curiosity, by contrast, can be actively promoted in the classroom: not by demanding attention, but rather by framing knowledge in terms of digestible information gaps. The simplest way to do so, familiar to teachers everywhere, is by asking questions of your students. There's a reason why the practice of teaching by posing questions—the so-called Socratic method—has stuck around for millennia. When it works, it makes learning feel as satisfying as eating or drinking.

—

Or even consuming harder stuff. One thing that's so pernicious about drugs that affect the reward centers of the brain is how quickly we can develop associations between those drugs and the sensation of reward.

Fortunately, the same applies to the fulfillment of curiosity. I sometimes tell this story to my daughter, and she doesn't always believe me, but it's true: On the oil platform, I developed such a strong association between a sense of reward and the attainment of useful knowledge that I felt, for the first time in my life, compelled to learn not by forces outside me, but from within. By the time I stepped off that platform, my brain felt *hungry*.

Actually, to be technical, what I stepped off was an icy step on the platform's exterior. I twisted my knee badly, and then it was back to the mainland for me. I recall that long helicopter ride vividly for two reasons. One: Immediately upon donning my protective dry suit and strapping in, I regretted drinking water beforehand. And two: It was on that flight that I realized that I had to go back to school to pursue an academic career. I needed to learn more—and faster.

That was the career trajectory that eventually took me to MIT, where I would get the chance to continue to learn alongside some of the best minds in my field, mechanical engineering—and also to learn about learning itself, thanks to my close proximity to researchers like Kanwisher, Gabrieli, and others working in the high-rise of cognitive science disciplines.

Systems-level cognitive scientists, in their quest to uncover how, where, and in what circumstances memories are stored, have uncovered deep processes that are utterly crucial to the job of every student. These findings fly in the face of the all-too-common idea that learning should be a struggle: an ordeal to be surmounted by force of will. In fact, whether a student is struggling mightily to read despite dyslexia, or struggling to keep her eyes open in a boring lecture, effort alone is often not enough to win her the day. For even highly motivated learners to reach their potential, we have to find ways to tally the practice of education with the demands of the systems-level brain. In many cases, the most important thing to do, practically speaking, is to find ways to promote curiosity.

Any veteran teacher can quickly rattle off strategies to promote curiosity that work even within the constraints imposed by the traditional classroom format. Some swear by connecting the curriculum to students' existing interests. (This approach can easily misfire, however, as with the tragically unhip English teacher satirized in the classic *Onion* op-ed, "Shakespeare Was, Like, the Ultimate Rapper.") Other strategies involve creating information gaps by means of the Socratic method, or encouraging students to skim a chapter before reading it in earnest, thereby setting up questions that only a careful perusal will answer. Still other tactics lean into the learning-progress theory of curiosity by, for instance, calling out common misconceptions about a topic before diving in.

But such strategies come with intrinsic limits. Most important, they only work well for motivated students. And then even if students are motivated in a general sense like I was as an undergrad (I wanted to learn but couldn't make myself curious), promoting curiosity can present a chicken-and-egg problem, requiring a little knowledge up front that, though tiny, can still be hard to choke down.

Is there a more organic way to lead students to drink from the fountain of knowledge—not through main force, but out of genuine thirst?

Kanwisher was discussing how MIT's Building 46 wasn't a traditional academic monolith so much as a bridge between levels of resolution when she stopped to make an important point. "The other piece not to be left out," she said, is the suite of labs on the McGovern side of the building where inquiry above the systems level takes place. "That is the part of the department where people actually think about thinking," she said. Although the systems level of neuroscience has revealed much about the mechanics of memory and the states that enable it, there's far more to learning than memory alone. And to understand more, I need to climb up higher: to talk to MIT's cognitive psychologists.

- IV -

LAYER THREE: REVOLUTION

Brandon McKenzie plowed through the water, pensive. It was midway through the semester, the final robotics competition for Course 2.007 was already looming in the calendar, and he was falling behind. Even now, on a weekday afternoon, Z Liang, Amy Fang, Alex Hattori, and others were probably honing their designs. Brandon, meanwhile, was swimming laps: always moving forward, never seeming to get anywhere.

His brain didn't switch off when he swam, however. The backstroke, in particular, allowed him to gaze skyward, watch the natatorium's rafters and ventilation ducts slide by, and think about whatever was bothering him.

Today, he couldn't decide what was bugging him more, his robot or his professor, Amos Winter, so he considered both in equal measure. Earlier in the week he had crossed the first major milestone en route to the final competition, but he had not done so with flying colors. He and his classmates had been tasked with demonstrating their "most critical module" (MCM, in engineer-speak), the component most directly responsible for earning points in the competition. This year's competition would be *Star Wars*–themed, prosecuted on and around enormous X-wing fighters. The team of instructors at MIT's Pappalardo Lab had outdone themselves this year and unveiled two identical perfect wooden replicas of the starships, twelve feet from wingtip to wingtip and ten from nose to stern.

It is possible, however unlikely, that you've never seen a *Star Wars* film and so don't know an X-wing from a TIE interceptor. An

X-wing, unlike some of *Star Wars*' stranger spaceships, is recognizably similar to a real-life, single-occupancy fighter jet, with the major exception that each of its wings can be made to split open like the covers of a book, presumably for the sake of combat maneuverability, to form an overall shape that, from the front, looks like a squat letter *X*.

The competition would unfold two students per X-wing. Brandon and his future adversary would line up on either side of the fighter's nose cone, their backs to the auditorium's audience, with their robots resting at knee height on a black wooden stage that also supported the X-wing. The starting bell would sound, then their robots, operating either under remote control or autonomously, would race to score points by completing any of a variety of tasks on or underneath the wooden starship. The most obvious way was to spin the X-wing's "thrusters," tube-shaped objects affixed to each of the starcraft's four wings, tucked in close to the central cockpit. In the *Star Wars* films, the thrusters function like rocket engines, but in this case, they more closely resembled spinning jet engines, with a thick wooden wheel mounted on either end of each thruster, designed to turn in place. Because there was one thruster per wing and two wings on either side of the X-wing, each student would have a crack at spinning both a lower thruster and an upper one. The faster their robots spun a thruster, the more points they would earn, with extra points awarded for the upper one, since it was harder to reach.

To complicate matters, a twenty-five-pound, cast-iron flywheel was built into each thruster, which would require serious torque to budge. The motors provided to 2.007 students, meanwhile, were either high-torque servo-motors, which would turn only slowly, or high-speed, lower-torque drive motors, not exactly up to the task of spinning a twenty-five-pound weight. Some students had begun building gearbox transmissions to solve the torque problem, a challenging prospect Brandon didn't relish, because even if he figured it out, there was still the challenge of manipulating the thrusters. Two schools of thought had emerged: You could either drive a wheel somewhere on the surface of the thruster, turning it by means of friction, or else you could insert a fork of some kind into the four "buttonholes" sunk into the thruster's flat face, and spin it as though gathering up spaghetti. Already, Brandon's classmate Z had been observed tinkering with a two-pronged fork in the lab.

If spinning the lower thruster was complicated, the upper thruster was especially forbidding, requiring a robot to either extend up an arm of some kind for a frontal approach or execute a hair-raising drive *behind* the X-wing, up a motorized elevator, and onto a shelf behind the starcraft, where it could approach the thruster from the rear.

Brandon had set his mind on what seemed to be a simpler plan. In addition to the thrusters, there were several other ways to earn points. Protruding from the ground in front of the starship's twin wings was a small acrylic stand, and hanging beneath it from their magnetized hands were three heavy, metallic toy bad guys: stormtroopers. Robots could earn points by removing these stormtroopers from their stand and magnetizing them to metal strips strategically placed throughout the game board, such as on the underside of the lower wings of the X-wing.

Brandon had consulted with Josh Graves, his roommate and right-hand man, and they had decided to go after the stormtroopers with a pincer-like device. That raised problems of its own, however, such as translating the circular motion of a motor into the lateral movement of the pincers. Brandon had cut sheet metal into six long strips and attached them to form three sets of blunt scissors, which a motor could close by pulling on a length of fishing line. He'd been excited to show Winter the progress he had made on this device, which was now officially his MCM.

Winter's plan for the day was to wander around the lab, spending about ten minutes with each student as they demonstrated their work. Brandon was adjusting his scissors at one of the X-wings, trying to get the device to latch on to a stormtrooper. On the other side of the X-wing, his classmate Zooey Bornhorst was fiddling with her own MCM. Winter came over to her first, and soon Brandon was overhearing sounds of praise. Like Brandon, Zooey was going for the stormtroopers, but her MCM had no moving parts. Instead, she'd figured out a clever way to slide a ferrous finger of metal next to the stormtroopers' magnets in such a way that pulled them away cleanly.

Winter walked over to Brandon's side of the X-wing, where he was still struggling to get a solid enough grip with his pincers to overcome the magnets holding the stormtroopers in place. Winter's expression was hard to read, but the stream of praise that had flowed so effusively minutes ago had dwindled to a trickle. Winter provided some advice: It would probably be necessary to cut a hook shape into Brandon's scissors, so they could act less like scissors and more like insect mandibles—but even then, Brandon would need not only to latch on to each stormtrooper, but also to pull it down in order to free up the magnet. "That," said Winter, frowning, "adds a layer of complexity."

Winter moved on down the line, leaving Brandon with his pincers. Amy Fang had perhaps the flashiest-looking MCM: a giant wheel meant to rotate the thruster by spinning against its flat face. Jordan Malone, the computer-aided-design wizard, had built an industrial-style scissor lift out of laser-cut acrylic parts. The day's biggest surprise, however, had come from Z, giggling off to one side. He'd adopted the spinning-fork method of thruster rotation, but instead of messing with a gearbox, he'd simply plugged the faster of the lab-provided motors straight into the most powerful battery available—the probably-not-explosive lithium polymer battery—rather than drawing power from the Arduino, which would have made the motor easier to control but produced a relatively staid output voltage. Now he held his fork up to the holes in one of the thrusters and turned it on. In a matter of seconds, the thruster was rotating at an astonishing twenty-five radians per second. Suddenly, Z was looking at the possibility of earning a huge number of points—far more than Brandon or Zooey could with their stormtrooper strategy, or nearby Ananya, whose arch-shaped robot was struggling to lift a block modeled after Han Solo's carbonite prison from *The Empire Strikes Back*. Z, it seemed, had discovered what appeared to be a loophole in the competition, and could barely contain his glee.

Now, a few days later, Brandon swam and thought. He'd met with Winter at his office, filled with vintage Transformers toys, to discuss strategies. It had been a confusing experience. Every time Brandon brought up problems he was having with his scissors, Winter always came up with a helpful suggestion, but at the same time, he always seemed to stop short of offering his true opinion on the overall design. "I felt as if he was just throwing me a bunch of more complex ways to fix my machine that was just kind of—it was discouraging, because it felt as if it would be a complete time sink," Brandon later recalled. He let out a sigh. If Winter did hate his design, he was keeping his feelings to himself.

THE INERT KNOWLEDGE PROBLEM

Brandon's suspicions were correct: Winter harbored little love for Brandon's scissor grabber. "That idea had tons of uncertainty," Winter recalled later, after the final competition. The mechanisms involved were "really kludgy, and weren't going to work well." But even if the scissors were simultaneously ineffective and overcomplicated, Winter couldn't just go and tell Brandon to give up on them and try something else. The moment he did—the moment he rattled off the approaches that would have worked better, which were manifold—he would have defeated one of the main purposes of the course: to teach students not just the principles of calculus and physics and mechanical engineering, but how to think *using* those principles. To bring inert lumps of knowledge roaring to life.

It still makes me cringe to think of all the time I spent as a student, building up piles of inert facts in my head with no intention of ever putting them to use once I'd puked them out onto a final exam. But my emotions pale compared to those of my close friend, the physicist and MIT educator Sanjoy Mahajan, who has made it his personal mission to revivify inert knowledge, especially in math and physics, wherever he finds it. To illustrate the problems with traditional rote learning, he frequently gives an example from Newtonian dynamics, one of his specialties as a physicist. He asks his audience to imagine a steel ball falling onto a steel table and bouncing back up. Neglecting air resistance, what, he asks the crowd, are the forces acting on the ball at the split second when it's touching the table, before it bounces? Most students who have taken a Newtonian physics course will claim that in addition to the weight of the ball, the table is also exerting some kind of upward-pointing, "normal" force on the ball. Usually, they say that this force is equal to the weight of the ball, since the ball is momentarily moving neither downward nor upward.

Mahajan rolled his eyes over a cup of decaf tea at the Atomic Bean Cafe, a hip, brick-walled coffee joint halfway between MIT and Harvard that has the Kinks and the Replacements on constant, loud rotation. Raising his voice above the music, he explained how students at top university after top university get it wrong. "I've asked this question at Olin"—the small, prestigious engineering college in Needham, Massachusetts—"I've asked this question at MIT, I've asked it of the students in Cambridge," he said, meaning the one in the UK, "and ninety percent will say the upward force equals the weight."

It's easy enough to prove that wrong. In his presentations, Mahajan invites a volunteer to place her hand on a table, palm down. Mahajan then reaches into his bag and pulls out a tennis-ball-sized rock. First, he places it gently on the back of his volunteer's hand and asks if it hurts: it doesn't. Then he picks it up and dangles it portentously, high above the volunteer's hand—which the volunteer always quickly withdraws in fear. Why, Mahajan asks, did you move your hand? After all, 90 percent of physics students said that the only forces acting on a steel ball (or rock) at the bottom of a downward plunge were the weight of the ball and the normal force from below—the same as when the rock was resting peacefully on the back of your hand.

The real problem, of course, isn't with the volunteer so much as with the vast majority of physics students. Everyone understands at a gut level that a falling object exerts more downward force than an object at rest—several thousand times as much, easily—which is why volunteers flinch when threatened with falling rocks. Advanced, successful physics students, however, routinely fail to connect that intuitive knowledge with the organized, pre-abstracted knowledge they've learned in their physics courses. It's almost as though the act of accessing memories created in the classroom shuts down the experiential part of their brain. To co-opt Woodie Flowers's phrase, they have learned physics, but they have not learned to think using physics.

The issue of inert knowledge crops up across courses of study. The question of how to coax third- and fourth-year humanities undergrads to apply strategies learned in their first-year writing courses, for instance, has become a stand-alone subfield of educational research.

And in physics, professorial worries about inert knowledge go back at least fifty years. In 1969, J. W. Warren, a London physics professor, presented 148 of his students with a series of questions that should have been no sweat, but which were structured more holistically than the students were accustomed to. For instance: Instead of asking students to puzzle out the magnitude of force x acting on a car driving in a circle at such-and-such speed, he asked them simply to diagram all the forces acting on the car, and to identify which force corresponded to the friction exerted by the ground on the car's tires. Less than a third of students were able to identify the forces acting on the car, and only *three* of them successfully diagrammed the direction of the frictional force, which should have pointed toward the center of the circle. (Without such a force, the car would simply slide away, as though on ice.)

Nearly half a century later, a group of like-minded professors, including Mahajan and me, posed the same exact set of questions to students of dynamics at MIT and the Olin College of Engineering. I wish I could tell you the modern students performed better, but ours made the same mistakes that Warren's did in the 1960s, at essentially the same rates.

If we had failed to fix the inert knowledge problem since the 1960s, it wasn't for a lack of trying. Western education has changed, at least on a surface level, considerably since Warren's study. The late sixties were boom times for both those seeking to challenge authority and also those convinced the West was falling behind the Soviets in science education, and Warren's study only added momentum to a well-intentioned if fractious push to shake up traditional classroom practices. In American teacher-ed schools, this drive often took the form of freewheeling, Dewey-eyed visions of education that centered more on students' professed interests than on a strict curriculum, and tended to involve more group work, less reliance on textbooks, a flexible daily schedule, and, where possible, pacing attuned to each individual student. Depending on which of these facets took priority, practitioners variously adopted names like *informal*, *child-based*, *indirect*, *project-oriented*, and *constructivist* to describe their approaches. Despite their internal differences, it will make sense for our purposes to group these systems together as "outside-in" approaches, aligned with Dewey's outside-in approach to conducting science. These sorts of systems leaned toward treating both curricula and the learning process as inviolable wholes, preferring not to over-engineer instruction, but rather let the student's natural inquisitiveness lead the way. Perhaps the most extreme form that outside-in pedagogy has taken is *discovery learning*, wherein students, acting like junior Newtons and Mendels, are tasked with discovering fundamental truths about the world for themselves.

Whether a given school adopted such changes in any significant way depended on a variety of factors. Some rewrote their mission statements with the zeal of converts; others, if my grade-school experience in India was any indication, remained wax-eared to the larger movement. At MIT, the trend inspired both research and instruction, including the revamped and revitalized Course 2.007 (originally called Course 2.70), to which, in 1970, Woodie Flowers added its now-famous robotics competition. Such a hands-on challenge, the thinking went, would defibrillate inert knowledge to life.

Outside-in critics of traditional education, Mahajan explained, had argued that " 'Well, students can't *do*.' " He sipped his tea. "And they're correct." But *doing* alone, in the outside-in line of thinking, wasn't sufficient as a remedy: The teacher couldn't just deliver step-by-step instructions for how to design and execute a complex project, because then students would only ever learn to paint by numbers. Rather, students had to carry over abstract knowledge into real-world practice for themselves. It was for this reason that, although Winter hinted to Brandon how he might improve his scissor module, he refrained from telling him to scrap it in favor of a better strategy.

Of such an approach, "I think that maybe the fundamental axiom is the teacher is a 'guide on the side,' not a 'sage on the stage,' " Mahajan said.

Even as such strategies spread, however, a countervailing group of educators began to voice reservations. Some drawbacks were obvious: hands-on learning could be resource intensive, for instance, and when schools eschewed such quantitative measures as grade point averages, student progress could prove hard to track. But there was also a deeper issue—an issue around which critics soon coalesced, convinced that the supposed cure for traditional instruction was worse than the disease.

Mahajan counts himself among them. "I think discovery learning," he said—outside-in instruction at its most radical—"is complete rubbish."

The fissure between the laissez-faire, outside-in camp and their opponents comes down to how the two sides interpret research at a specific level in the cognitive high-rise—the level at which we've just now arrived.

By the late 1950s and 1960s, the gap separating the uppermost levels of brain science and the lowest levels of psychology were in some instances beginning to close. This process was, and remains, fraught with tension—perhaps unavoidably. The research purview of systems-level brain scientists, though mind-bogglingly complex, consists essentially of things, not people: neurons, memory, drive states, and so on. When you zoom your microscope of inquiry out, however, until you're operating at a *psychological* level, something ineffable slips into your field of view. Suddenly, your jurisdiction includes not just biological machinery but entire human beings in all their glory—and with them, all the unanswered head-scratchers about the human experience that have vexed serious thinkers for millennia. Closing the gap between brain and psychological sciences, then, involves sidling up to scorchingly hot philosophical questions: the nature of human volition, for instance, and even the nature of consciousness itself.

More immediately relevant to education: The borderline where brain science meets psychology is also where researchers start talking less about *memory* and more about *learning*. The two words are not synonymous. Memory entails the mere storage of information, while learning involves abstracting meaningful rules and patterns from that information and putting them to work. Although learning is certainly possible at lower levels in the cognitive high-rise—the brain's letterbox, for instance, learns to abstract the identity of letters from their shapes—it is at the psychological levels that actionable learning strategies for students and teachers begin to pop up in earnest.

Both outside-in- and inside-out-leaning education reformers agree that the way most classrooms today package and deliver information isn't working; it results in inert bodies of knowledge. They also concur that a more user-friendly, less cognitively obstructive approach to instruction is the key to fixing this problem. Where they disagree, however, is in how they define "user-friendly"—a point of ambiguity that arises at the charged point in the cognitive high-rise where neural activity gives way to thought, and memory gives way to learning. The result is two separate plans for overhauling education that, at least at first glance, cannot possibly coexist.

THE COGNITION SWITCH

The 1950s had been a golden era for the behaviorists, with their allergy to any mental process between "input" and "output," but by the end of the decade, change was in the wind. Across a variety of fields, researchers had begun prying apart some of the mechanisms supporting how we think. Across brain science and *cognitive psychology*—that is, the subfield concerned with the inner workings of thought—and with help from sibling fields like linguistics and computer science, a loose collection of trailblazing scientists kicked off a new phase in how we think about thinking, later known as the *cognitive revolution*.

In terms of education practice, it was the psychological side of the revolution that broke through first. The genesis of cognitive psychology was closely tied to the American rediscovery, in the 1950s, of Jean Piaget, a Swiss psychologist some two decades Thorndike's junior whose ideas had been all but forgotten in the English-speaking world.

Today, Piaget is rightly esteemed as one of the titans of twentieth-century psychology, whose long career not only paved the way for contemporary, outside-in educational practices but also brought the research field of child development into the modern scientific era. His early career is also notable for another reason: It's a testament to how different minds can respond differently to the same stimulus. To E. L. Thorndike and his allies, the first modern intelligence test, developed by the French psychologist Alfred Binet in 1905, presented an opportunity to sort learners by their supposedly innate ability. To Piaget, however, it was a chance to identify not what separated individuals, but rather what we all had in common.

Piaget grew up at the turn of the twentieth century in the Neuchâtel region of French-speaking Switzerland, where he exhibited a precocious interest in zoology in general and mollusks in particular. He zoomed through his coursework so quickly, in fact, that by age twenty-one, he had already earned his PhD but only begun to find himself. Shortly after delivering his dissertation, he began a new career in the still-young field of psychology. He studied in Zurich under the famed psychoanalyst Carl Jung and psychiatric pathologist Eugen Bleuler, and then accepted a teaching position at the Sorbonne in Paris. There, he made the acquaintance of Théodore Simon, the longtime collaborator of Binet, who had died years earlier.

Soon, Piaget was in the intelligence-testing business. Charged with standardizing for children a test originally designed for adults, his job was simply to administer the test orally to the young test takers and record the number of incorrect answers. He strayed beyond his brief, however, and in an attempt to understand the children's reasoning, followed up their incorrect answers with penetrating questions of his own, a method that would become a staple of his long research career.

Children, he discovered, appeared to make the same sorts of mistakes at roughly the same ages, which unfolded in a somewhat predictable progression. This suggested the existence of discrete stages of development that children passed through as they grew up. The details of these stages, on which Piaget bestowed names like "sensorimotor" and "preoperational," have not held up to modern scrutiny. His underlying assumptions, however, have not only remained mostly unassailable, but have inspired generations of further research.

Informing virtually all of his subsequent work was an observation he made at the very start of his career. Young children, he discovered, only make childlike mistakes as a result of naïveté, not an underdeveloped ability to reason. For instance, he realized, young kids commonly assumed that, say, a dozen pebbles in a pile would magically increase in number if the pile were spread out over a larger area. When they discovered that the number of pebbles actually stayed constant, they often experienced it as a revelation.

In this case and others like it, the most important part of the learning process was that moment of realization, which Piaget viewed not as an act of data storage but rather one of invention: the active creation of a personal rule of thumb about the world. What if, Piaget reasoned, this sort of creative process is actually how we assemble the sum total of *everything we know*? In addition to fundamental rules of the sort illustrated by the pebbles, this theory appeared to explain learning via simple observation as well. When you think back on a visual scene—say, your breakfast this morning—it will probably appear in your mind less like a perfect photograph and more like an impressionistic painting, with only a few relevant details fleshed out. (You might, for instance, recall that you ate Cheerios, but not how many.) This selective attention to detail makes perfect sense if you think of learning not as a sequence of passive snapshots but rather as an active process of fabrication, adding up to your own personal homegrown model of reality, in which certain facts are necessarily more important than others. Such a vision of learning could even explain—and this was something a later generation of educators would seize upon with zeal—how some people appeared to improve not just their bank of knowledge but also their problem-solving and decision-making *skills* over time, as they built up representations of raw knowledge as well as personal theories about how best to act on it.

The sum total of your knowledge, in this model, consisted of innumerable individual sketches and rules—what Piaget referred to as *schemata*, or *schema* in the singular—assembled into a treelike whole, with long branches growing out of a trunk of basic knowledge.

Prior to the late 1950s, Piaget wasn't so much ignored in the U.S. as politely entertained—he gave a handful of Ivy League lectures in the 1920s and 1930s—and then forgotten in a miasma of behaviorism and disrespect for the supposedly feminine field of child development.

Then, on October 4, 1957, a tiny white dot passed over the American night, fraught with implications for the United States' military security, not to mention its national ego. Paying especially close attention to Sputnik and its policy aftermath was the MIT physicist Jerrold Zacharias, a Manhattan Project alumnus whose work in the field of nuclear magnetic resonance would eventually lead to the development of the MRI. A year earlier, Zacharias had secured federal funding to try to find a way to close the perceived "knowledge gap" separating American students from their Soviet contemporaries. Now, under the roving eye of Sputnik, such efforts took on new urgency. President Eisenhower quintupled the National Science Foundation's budget, appointed MIT president James Killian as his first special assistant for science and technology, and created the President's Science Advisory Committee, on which Zacharias sat.

To Zacharias, the influence of behaviorism over U.S. education was a major cause of the knowledge gap. In an effort to find a new way to motivate students, he hired a close personal friend, the Harvard psychologist Jerome Bruner, whose skepticism of behaviorist thought was so pronounced that, a year earlier, he had led a psych-department splinter group away from the oversight of the uber-behaviorist B. F. Skinner.

In 1959, Zacharias convened a conference of psychologists and research scientists in Woods Hole, Massachusetts, to discuss how to apply to science education the new findings coming out of the cognitive sciences. Bruner turned his conference report into a book, *The Process of Education*. Unlike most American psychologists, he had been a strong believer in Piaget's work for years, and included a discussion of Piaget's alternative to behaviorist models of memory. The book proved a surprise success, selling nearly half a million copies in four years.

Just like that, Piaget was back on the map, and this time, America was ready. By 1960, the cognitive revolution was in full bloom, and its epicenter was Cambridge, Massachusetts. If you'd stood at the eventual site of the Atomic Bean Cafe, midway between MIT and Harvard, you couldn't have heaved a sidewalk brick without striking a cognitivist breaking new ground. At Harvard, these included Bruner; his colleague George Miller, who, as I'll explain ahead, had begun to plumb the depths (more like shallows) of working memory; and the psychologist Roger Brown, who was studying language development in children. At MIT, meanwhile, you could find computer scientists beginning to think through how to build a mind from scratch, as well as, crucially, the cognitive linguist Noam Chomsky, whose scathing 1959 review of Skinner's book *Verbal Behavior* served as one of the cognitive revolution's opening shots. Into such a heady cognitivist milieu, Piaget's theories fit like a missing puzzle piece.

Mahajan leaned back and glanced around the Atomic Bean. "Representations have to be built in a person's cognition and mind. That part, I think, is unarguable," he said. The problems with

Piaget's theories of learning, he said, only arise in the transition from theory to practice: when one seeks to turn them into "a philosophy of education."

The Piaget-inspired pedagogical strategy destined for greatest controversy was the aforementioned discovery learning, which assumed that the best way for a learner to add new knowledge to her schematic tree was not to have it served up by a teacher, but rather to discover it organically. Teachers, in this model, served as facilitators of discovery, their sole instructional duty to teach learning and problem-solving strategies, not actual facts about the world. The content of school thus became a means to the end of teaching students to "learn to learn," as a newly popular maxim went.

CAMPUS CONSTRUCTION

Teacher-ed programs served as the main, albeit circuitous, route by which Piaget's ideas found their way into American educational practice. Their channel into MIT was more direct, however, and so they made their stamp here more immediately. And in its way, MIT soon stamped back.

Piaget's main ideological vector on campus was one Seymour Papert, a South African mathematician and computer scientist who spent four years working with Piaget in Paris before making his way to MIT. Peering past the massive institutional computers he was using in the late 1960s, Papert saw ahead to a time when computers would be cheap and ubiquitous enough that children might want to learn to code. He developed Logo, a simplified coding language, to help them do so. "Some think of using the computer to program the kid; others think of using the kid to program the computer," he wrote in 1971, in a prod perhaps aimed at Skinner, who was still proselytizing his teaching machines a couple miles up the Charles River. Papert developed the Logo Turtle, an early robot housed in a transparent dome the size of a halved basketball, to help students turn the esoteric experience of programming into something tangible. The Turtle, which predated ubiquitous computer video displays, *was* the display: it could drive on top of a large sheet of paper, drawing geometric designs as it moved. For students lucky enough to have access to one, the purpose of mathematics changed; it went from a problem-solving tool to a creative tool.

Today, Papert's legacy at MIT continues, borne forward into the smartphone age by his student and longtime collaborator, Mitchel Resnick.

Resnick stands about six-foot-four and has worn a salt-and-pepper beard for as long as I've known him. When he talks to you, he leans back in his seat and looks through the lower half of his eyeglasses with twinkling eyes. He leads MIT's Lifelong Kindergarten group, which is dedicated to "designing, creating, experimenting, and exploring" well into adulthood.

Early one morning in his warm, primary-color-filled office, before the rest of his team had shown up for work, he examined the legacy of his intellectual forebears. "Piaget—I think one of his big contributions was that learners, kids in particular, actively *construct* knowledge. It's not fed to them," Resnick explained. "And Seymour totally agreed with that, but he said, 'And one of the most effective ways of people constructing knowledge is through constructing things in the world.' "

Seymour Papert meant that quite literally, in fact. To his mind, the best way to erect a flowering tree of knowledge was to build objects: in the classroom, in the garage, in a computer. "The construction that takes place 'in the head' often happens especially felicitously when it is supported by construction of a more public sort 'in the world'—a sand castle or a cake, a Lego house or a corporation, a computer program, a poem, or a theory of the universe," Papert wrote in his book *The Children's Machine*.

"The act of creating things in the world sparks me to think about things in a different way," elaborated Resnick. "And as I think about things in a different way, it sparks me to create new sorts of things."

Today, *constructionism*, as Papert's philosophy is known, has a reputation for applying mainly to STEM (Science, Technology, Engineering, and Mathematics) fields, particularly engineering; but Papert, who died in 2016, would have disagreed with such a limited reading of his idea. In fact, the humanities have long relied on constructionist teaching methods; if anything, the STEM fields are just now catching up, borrowing from the greatest hits of humanities education. It's hard to pass a course in the creative arts, for instance, without committing an act of creation: a painting, say, or a musical composition, or a poem. Meanwhile, other fields of study in the humanities, particularly at the college level, are built around prose writing, an undeniably creative endeavor.

"The act of writing helps you to organize, express, and share your ideas," said Resnick. Indeed, he said, "We want everyone to learn to write—it's not only for people who are going to grow up to become professional writers." Not only does writing help students learn to formulate and express their ideas logically, but it also has, at least in theory, a virtuous social effect. "I think it's important to help give everyone a voice," he said, "to make them feel like they can be an important contributor to the ongoing things in the society."

Today, however, many of the visible and hidden rails that guide our lives are written not in English or Spanish or Mandarin, but in the language of computers. "For me, learning to code is very similar to learning to write," said Resnick. "It's not just for people who are going to use it in their jobs, and it's not just for learning computational concepts, although it's good for that. It's another way of expressing your ideas and to organize your thinking."

When Papert made the argument that everyone should learn to code in his 1980 book, *Mindstorms*, the idea caught on quickly, and Logo coding was even made a mandatory subject in UK and Costa Rican public schools. However, despite its popularity and Papert's fervent hopes, the programming language didn't revolutionize education. Logo itself was part of the problem. Papert intended the language to be easy to learn, but its syntax was still reasonably difficult, which turned off beginner students and instructors. As a result, even as computers began to pop up in classrooms and computer labs in the 1980s and 1990s, coding literacy failed to take off. Instead, a different sort of computerized diversion gathered steam: educational computer games, most of which functioned as glorified content-delivery systems not so different from Skinner's teaching machines.

"Little by little the subversive features of the computer were eroded away," Papert wrote. "The computer was now used to reinforce School's ways."

The promise and shortcomings of Logo stuck with Resnick, however, who earned a computer science PhD at MIT under Papert and soon began conducting research of his own. Resnick began by tinkering with Logo, taking it in new directions that included a hit retail product, LEGO Logo, which allowed students to control Lego robots using Logo programming. It was highly engaging—but any disinterested observer would have to admit that the Lego side of things was more fun than the programming. That sparked an insight in Resnick: Perhaps the rewarding, intuitive act of snapping together multicolored blocks could itself be replicated on the programming end.

The programming language he and his team developed to make this possible is called Scratch. Instead of a finicky text-based language, Scratch allows users to code using simple, block-shaped visual elements. Best of all, those blocks of code are set up, as a default, to control the movements of characters and objects in animated videos, meaning that brand-new users can walk away from their first Scratch coding session with a funny cartoon under their belt—typically, of a cat dancing across the screen. As of this writing, Scratch is within striking distance of the scale Papert originally envisioned for Logo, with 30 million registered users on its website. Perhaps the program's most important endorsement, however, is that learners use it not just when assigned, but also in their free time—my daughter included—and often in pairs and teams. "I sometimes say that the ultimate goals of what we're trying to do in all of our projects is help people learn to think creatively, reason systematically, and work collaboratively," Resnick said.

These sorts of aspirations might seem uncontroversial, but beneath them still runs the cold current of Seymour Papert's more subversive ambition: to change our inherited rules of education. One thing that bothers Resnick in particular about the Thorndikean status quo at school is its reliance on letter and number grades for the purposes of sorting students. "I would certainly be aligned with reducing the importance of grades," he said. "At MIT, as you might know, the first semester there's not grades. And I would extend that further." At the Institute, most of the faculty acknowledges the grade-free initial semester as a good thing, a mental health boon for students at a delicate phase of their development. There is still plenty of internal pushback, however. Resnick said that one physics professor who teaches juniors told him that because first-year students were deprived of the motivating force of grades, they eventually turned up underprepared for his class. Resnick thought that he was framing the issue the wrong way. "If there's an MIT student who isn't paying attention and trying to learn things freshman year because there aren't grades," Resnick said, growing grave, "that student shouldn't be here. That's a real mistake that MIT made, to admit a student who's only motivated because of grades."

The idea belongs to a time-honored line of outside-in thought. Behind every number or letter standing in for a student's potential lurks an undeniable, tautological flaw: Put simply, such metrics don't measure what they don't measure. Consequently, it's impossible to know how many promising students exist around the world whose grades and test scores only obscure what they're really like as learners. Or worse, perhaps grades, test scores, and the classrooms that seem to revolve around them actively *prevent* certain students from getting excited about learning. A broad shift toward outside-in, project-based instruction, including MIT's homegrown brand of constructionism, might offer students new ways to prove themselves, while giving them the chance to discover their true passions. Only then, Resnick said, "will they be willing to put in the hard work that's necessary to be really successful."

Adopting constructionism and its fellow outside-in approaches, in short, seems like a no-brainer: They provide a welcome injection of motivation, contextualization, and activated knowledge into an education system in dire need of all three—plus, as a bonus, a more forgiving way of thinking about the educational winnower. There's just one problem: The murmurs from the growing and increasingly influential camp of education researchers who insist that outside-in strategies do not support the cognitive processes necessary for learning, and may even smother them.

INSIDE-OUT THINKING STRIKES BACK

To talk to advocates of outside-in education like Resnick, the work of converting their ideas into classroom practice is still only beginning. (After all, grades still exist.) To hear inside-out thinkers like Mahajan speak, however, the exact opposite would seem to be true.

Yes, old-school, traditionalist teaching still prevails, but everywhere you look, outside-in reformers seem to be slipping their assumptions into the curriculum and, despite their good intentions, systematically and kindheartedly obstructing learning.

By way of an example, Mahajan brought up how his daughter was taught to read at a local Cambridge public school.

"All across America," he said, "kids are read to and given lists of words to memorize," a strategy that is meant to make learning to read as natural as learning to speak. This approach got its start in the "Whole Language" movement of the 1980s, which eschewed traditional, phonics-based strategies like breaking words down into small parts and "sounding them out." Rather, the thinking went, in the same way that children naturally map the sounds of vast numbers of words onto their meanings just by hearing and speaking them, perhaps they could replicate that feat with written words if given enough of them to look at. By the late 1980s, Whole Language was spreading like spilt milk.

It was, in retrospect, an unmitigated failure, for reasons that became evident only in the wake of the brain research discussed in the prior chapter. The brain, it turned out, is innately wired to recognize speech but not written language, and so it's far easier to learn the former through exposure alone. Evidence supporting phonics as the superior tactic began to mount as early as the mid-1980s, and in 1996 the Whole Language movement received a dollop of bad news that it could no longer ignore: California, which had gone all in on Whole Language, had become the worst-performing state on standardized reading tests.

"I realized it was a total disaster," Mahajan said. "My daughter," who had been part of a Whole Language–inspired reading program, "couldn't read 'dog.' She's trilingual! She couldn't read 'dog' at age seven."

For those looking to disprove the utility of discovery learning, the collapse of Whole Language provides a striking but narrow case. A broader line of criticism came in the form of a 1985 study out of New South Wales, coauthored by the educational psychologist John Sweller.

Here's a pop quiz: One group of students works at solving algebra problems. A second group doesn't solve these problems themselves, but instead studies "worked examples"—that is, algebra problems that have been solved correctly ahead of time, with the intermediary steps spelled out. When presented with a new set of algebra problems to solve, which group fares better?

Surprisingly enough, according to Sweller's study, it was the worked-example group who prevailed, despite having logged less time practicing problem solving. Similar results were replicated again and again in the 1980s and 1990s, mainly in math-related fields including statistics, geometry, and computer programming. These findings rankled outside-in partisans, who had long argued that classroom time and energy was better spent on problem-solving strategies than domain-specific minutiae. Perhaps more troubling, however, was the conclusion drawn by Sweller and his allies that general problem-solving skills are essentially unteachable, like learning to walk or speak one's native language. If true, all that time outside-in educators and their students had spent on complex challenges in hopes of developing general problem-solving abilities had been time wasted.

Or worse: Perhaps all that problem-solving practice had actively *impeded* learning. To find theoretical support for this idea, Sweller's camp had only to look back a few decades to one of the foundational studies of the early cognitive revolution, *The Magical Number Seven, Plus or Minus Two: Some Limits on Our Capacity for Processing Information*, by Harvard's George Miller. The study was the first modern probe into what became known as working memory or working attention, an extremely ephemeral category of memory that corresponds to essentially all the information you're consciously aware of at any given moment. Information temporarily stored in working memory has two possible fates: either longer-lasting storage in short-term memory (and possibly long-term memory down the road), or else quick forgetfulness. When they're fully concentrating, Miller determined, most people can recite back roughly seven "chunks" of data from their working memory—a "chunk" being a highly subjective thing. For instance, the number 2001 would occupy just a single chunk for any fan of Stanley Kubrick, while the number 7845, chosen at random, would demand a chunk for each digit.

In 1988, Sweller fired off an article that chilled the ventricles of the outside-in camp. What if, he wondered, the lackluster performance of problem-solving students could be explained by the ineluctable limitations of their working memory? According to *cognitive load theory*, as he dubbed it, the cognitive demands of thinking through *how* to solve problems occupied critical slots of working memory that could be put to better use. During the trickiest moments of studenthood, he reasoned, it would be better to have all cognitive hands on deck, not occupied working their way through unnecessary puzzles. "There seems to be no clear evidence that conventional problem solving is an efficient learning device and considerable evidence that it is not," he wrote.

The findings provoked a fractious debate that rages still in the pages of education and psychology journals. Only very recently, however, have neuroscientists begun to satisfactorily explain the mechanisms underpinning cognitive load.

To explain, let's take a trip back to MIT's Picower Institute. Back in the 1950s, Harvard's George Miller had suggested that human working memory could maintain as many as nine chunks of data, but a series of follow-ups had thrown cold water onto that number and shrunk it down to four. In the 2010s, the Picower researcher Mikael Lundqvist came up with a plausible insight as to why. Appropriately enough, he had a four-digit office number, which necessitated constant mental recitation on the elevator ride up to speak with him.

The fleeting nature of working memory may not be a limitation of cognition so much as a feature, Lundqvist explained. After all, you *want* the stuff of conscious thought to flow freely, not get stuck on points for long periods of time. The main question concerning the "how" of working memory, then, is what makes it so ephemeral. As in short-term and long-term memory, scientists believe working-memory representations to be encoded in assemblies of neurons, held together by patterns of strategically strengthened or weakened synapses. What, mechanistically speaking, allows working-memory cell assemblies to build up their synaptic patterns in mere moments and wipe them clean just as quickly? For a long time, scientists believed that your prefrontal cortex—the site of working memory—holds on to its representations by maintaining a consistent buzz of activity in the cell assemblies corresponding to the items in your thoughts (that is, right until your attention wavers, and the action shifts to a different group of neurons).

In 2011, Lundqvist was working on his PhD in his native Stockholm, using computers to model working memory, when he hit upon something new. "Instead of using consistent spiking, I used synaptic waves to store memories," he said. In such a model, the chunks preserved in one's working memory would be preserved not all at once, as previously thought, but one at a time.

Lundqvist has a muscular frame and shoulder-length blond hair, and with his Scandinavian accent, he could almost be confused for someone out of a Norse saga. Only his eyes give away the sleep-deprived research scientist in him. Since joining Earl K. Miller's lab at MIT, he has been putting in long hours to get his model right—and in a far trickier study system than computers. His data now come from the brains of a pair of monkeys, outfitted with arrays of electrodes that tell Lundqvist when tiny regions of the prefrontal cortex are active. This technique is more invasive than an fMRI, but fMRI scanners can only describe neural activity on a second-by-second basis, while the electrodes are accurate down to the level of milliseconds. A series of these monkey studies have now returned results supporting Lundqvist's burst-firing model.

The difference between a continuous model, with its incessant spiking, and the burst-firing he proposes may seem subtle; but the burst model provides the most enticing explanation yet for working memory's inherent limitations. Imagine if the four available chunks of working memory were a four-word message written in wet sand at the beach—SANJAY SARMA IS HERE—under constant threat of erasure by crashing waves. In the old model, every letter of those words is refreshed simultaneously, as if by magic. In Lundqvist's new model, by contrast, it's as though the only way to maintain the words is to retrace them bit by bit, with just a single index finger. If I grow greedy and try to add a fifth word to the end of the message (SANJAY SARMA IS HERE *TODAY*), I might not finish in time to rush back and preserve the first word from the onrushing waves. Soon enough, I'm back to a four-word message, albeit a new one: SARMA IS HERE TODAY. This theoretical "refresh rate" limitation, as Lundqvist's team refers to it, provides a strong explanation for why, every time you add a new chunk of information in your working memory, you're likely to lose an older one.

LEARNING, APPROXIMATELY

It is possible to plumb working-memory capacity using softer-edged metrics than chunks, such as the sheer length of a verbal passage someone can repeat, or the fidelity with which they can replicate an image. Using such strategies, researchers have shown working memory to be both highly variable and correlated to measures of general intelligence. One upshot is that whenever something clogs up working memory, your all-around problem-solving abilities take a hit. Certain disordered conditions, including schizophrenia and ADHD, can interfere with working memory. Relatively temporary intruders, too, can get in the way. One particularly pernicious example that I've touched upon occurs in stereotype threat, when distracting thoughts related to such factors as gender, race, and socioeconomic status eat into working memory and depress students' test scores.

The wrong sort of instruction design, the inside-out camp argues, can only magnify such issues. When you're working on a complex problem, Mahajan explained, "most of your memory is filled up with 'search'": the act of running through hypothetical courses of action, seeking the one that will lead to a desirable solution. "You think to yourself: 'Well, what do I know, where do I have to get? How far away am I?'"

Learning via worked examples, instead of solving problems for yourself, is one potential way past such working-memory roadblocks. The inside-out faction also endorses "overlearning" certain facts, like the multiplication tables, so that summoning up those facts during problem solving becomes undemanding.

Mahajan has a different trick. He is, it turns out, one of the world's best estimators. Once, after a colleague of his was issued a large, confidential settlement during some legal proceedings, a journalist asked Mahajan to estimate the size of the cash award. Mahajan's highly educated guess was so close to the actual number that the issuer decided someone must have spilled the beans, and sued.

Mahajan's feat should have come as no surprise, however, since he literally wrote the book on approximation (his preferred term of art), *Street-Fighting Mathematics*. The title gives a fair sense of his approach to solving complex problems: First do a quick-and-dirty run-through of the problem, complete with an approximate answer, and only then aim your full working memory at the actual task of solving it. This shoot-first-ask-questions-later approach works not just for STEM problems, he said, but even for writing. "You have to, by hook or crook, write a completely shitty first draft," he said. "It's going to be very, very approximate," but then, on your next pass, with more working memory freed up, you can make it intelligible.

One thing worth noting about Mahajan's problem-solving strategy is that it is, in fact, a problem-solving strategy, and therefore a departure from the most extreme inside-out position, which holds that education should consist solely of content, not process. What Mahajan teaches is process *as content:* a strategy for how to direct your thoughts that is not so different from how a coach advises an athlete to move her body.

In fact, the sometimes-imperfect control we wield over our thoughts appears to have real commonalities with our oversight of our bodies. The bursts that Lundqvist has detected moving through monkeys' prefrontal cortices are close, timing-wise, to the speed of the fastest possible actionable information flow in the brain—"on the border," he said, of "the speed of thought." And how might conscious thought control the contents of working memory? Here, too, Lundqvist's model holds tantalizing hints. We can't say how volition arises in the brain, or even if we truly have free will over our thoughts and actions. We do, however, know that the patterns of neural activity that occur when you create a working-memory representation are suspiciously similar to the patterns that arise when you decide to, say, move your arm.

The key is the important neural phenomenon of brain waves: electromagnetic peaks and troughs that correspond to the massed activity of large numbers of neurons working in concert. In the case of your control over your limbs, there is a sort of holding pattern, known as a beta wave, that prevails in the motor cortex when nothing much is happening. It appears to act like a wet blanket, inhibiting unwanted activity, and must be removed if you want to use your arm. With the wet-blanket beta waves out of the way, a series of high-frequency gamma waves arrive that correspond with arm movement. Curiously enough, these come in bursts remarkably similar to those Lundqvist has observed in his working-memory studies, which likewise appear to be held in check by a wet-blanket beta-wave state. The theoretical implications of such similarities are remarkable: the same volitional process that allows you to control your limbs may also grant you control over the flow of your thoughts.

"The dynamics and everything else looks very similar between the motor cortex in movements and the prefrontal cortex in abstract thoughts," he said. "But it's still open to debate."

I appreciated Lundqvist entertaining such questions, but perhaps what I appreciated most was how he was willing to fly near neuroscience's version of the sun: the question of volition itself.

"Where the volition actually is, is like the ultimate question, right?" he said. "Is it just due to the reinforcement learning—that we learn to recognize the right situation to do something, and it looks to us like it is volition? Or is it . . . ?"

He trailed off, perhaps feeling his wings grow warm.

—

The inside-out versus outside-in debate comes down to similarly hot topics. A book from 2009, titled *Constructivist Instruction: Success or Failure?*—the cover depicts a pair of debate podiums—valiantly attempts to give each side a fair chance to respond to one another. Instead it quickly devolves into both passive aggression and speculative arguments over whether problem solving is a skill that can be taught, and whether learners actually have volitional control over their working memory. Like geopolitical rivals claiming mineral rights all the way down to the center of the Earth, it can sometimes feel like the two sides are trying to split up the cognitive sciences beneath their feet.

It would almost be surprising, however, if, in our ascent up the cognitive science high-rise, ideology *didn't* rear its head at the level of cognitive psychology. Both sides want to make education better, but once your scope of study includes human beings in their messy entirety, you inevitably run into questions like *Better how? Better for whom?* Should students follow their own impulses in determining what to learn, or should they learn to take strict control of their own thoughts and stick to topics their instructors deem important? As we continue our journey upward—and then out into the wider world, where the cognitive science of learning gets applied—such fractures will persist: an unavoidable, unignorable feature of any future educational landscape.

But that's no reason to give up. In the wise words of Luke Skywalker, Jedi Master:

Breathe.

Just . . . Breathe.

Back in MIT's swimming pool, that was what Brandon was trying to do, even as visions of stormtroopers, Han Solo, and X-wing fighters floated unbidden through his working memory. *Star Wars*, he realized, had begun to take on stressful connotations that he wasn't sure he appreciated. The team began a new drill: backstroke flip-turns, the disorienting, high-speed technique swimmers use to change direction in a race. The plan was, as with those multiplication tables, to overtrain the maneuver until it became truly second nature, and Brandon found his mind pleasantly occupied by the activity, more cognitively demanding than laps. From the surface, the technique looks like a frenzy of activity, and yet, Brandon found himself in a state of clarity that had been eluding him all day.

Indeed, sometimes there is order to be found in seeming chaos. It's a lesson we all learn starting in infancy, as we strive to abstract meaning from the cacophonous sounds and sights and smells that surround us. And, appropriately enough, it's a lesson adults, too, can take away from the seemingly disorganized learning behaviors of babies and young children. No one on campus understands this better than Laura Schulz, a cognitive psychologist who studies the deceptively rigorous logic that youngsters employ to make sense of the world.

In the years since Bruner reintroduced Piaget to America, Schulz explained, his ideas have gone through some changes. "Piaget's stage theory, of course, collapsed," she said, leaving behind only its most foundational assumptions. In the mid-1980s, a new, more rigorous theory began to take its place, which likened children to tiny scientists, an idea John Dewey might have welcomed. As Schulz's doctoral advisor and frequent collaborator, Alison Gopnik, has written: "The basic idea is that children develop their everyday knowledge of the world by using the same cognitive devices that adults use in science . . . That is, they develop theories. These theories enable children to make predictions about new evidence, to interpret evidence, and to explain evidence." This idea, fittingly named the *theory theory*, provides a compelling explanation for how we incorporate new data into our existing trees of knowledge. When evidence comes in that warrants a modification to an existing theory, a learner must add to the original, change it, or, in some cases, jettison it altogether in favor of a new explanation. As Schulz and her collaborators have shown, when children encounter evidence that defies their personal theories (often arriving in the form of deceptively designed toys), a concerted exploratory session often follows, in which they try to suss out what's going on. This outcome fits hand in glove with the learning-progress theory of curiosity discussed in the prior chapter. Not only does surprising information trigger the drive state of curiosity, but also, at the level of conscious learning, it transforms passive observers into de facto scientists.

Often, however, perplexing new information doesn't come with readily discoverable explanations attached. Happily, our brains appear to have evolved to deal with such contingencies remarkably efficiently, starting in infancy. Babies, it turns out, interpret events using not just a parsimonious approach, but often the best *possible* approach given the limited data available to them, similar to what statisticians refer to as Bayesian reasoning.

In one study of the hundreds that support this work, Schulz and a graduate student systematically played with deliberately broken toys while babies watched, with their mothers close at hand. When you're a baby who wants to play with a toy that's not working, your world shrinks down to just two options: try to figure the toy out, or hand it to an adult. Depending on their observations of how Schulz and her research partner fared with the toys (Did it always work for one of them but not the other? Or did it sometimes fail to work for both of them?), babies strategically decided whether to experiment with the toys themselves, or pass them directly off to Mom. The upshot was remarkable: In the face of extremely sparse information, the babies routinely made the optimal choice. They were acting, then, not just like tiny scientists, but tiny *statisticians.* And although this innate skill may be harder to observe in the everyday life of adults, in part because the pile of statistical data we're constantly integrating is larger and more complex, there's no reason to believe we stop adding to our tree of knowledge using these same principles. Such statistical reasoning may, in fact, turn out to be the glue holding together the theory of mind Piaget proffered early in the twentieth century.

You might reasonably assume that, given such strong support for Piaget-aligned models of learning, Schulz would also endorse outside-in education practices, but here she steps back.

"I don't think we're there yet," she said. On one hand, "there's no way you could learn calculus by discovery," but on the other, "we want engineers who will build things," she explained, and that calls for a certain amount of learning by doing. As we ponder how to educate as many people as possible, as effectively as possible, it's clear that we'll need elements of both approaches. We need a wider tolerance for diverse student motivations and interests, as well as the powerful ability of outside-in pedagogy, when used correctly, to make the dusty skeleton of inert knowledge jump out of its coffin and dance. However, we also need to marshal our resources—money and time, yes, but also students' cognitive resources—to teach useful, hard knowledge. And so we ignore the potential benefits of inside-out instruction at our peril.

The field of education cannot wait for the perfect answer. We need to integrate multiple viewpoints in real time, and to update that mixture as new evidence rolls in. The best response to the fractured state of educational psychology, fittingly enough, is not to take sides—but it's not to seek a haphazard compromise, either. Rather, we must adopt a Bayesian-style approach. As a young child behaves like a scientist in constructing her model of the world, we scientists and educators must establish a working theory of what works best in educational practice, while remaining open to changing it as needed.

Hard at work in MIT's pool, Brandon McKenzie had been flipping, setting his feet, and launching off the concrete wall for what seemed like hours when he suddenly figured out what to do about his robot. Was he upside down or right side up when he changed his mind? Was his decision-making process truly volitional, or the product of environmental cues, or perhaps the result of a complex interplay between his working and long-term memory? Despite the remarkable advances now being made at all levels of cognitive science, it's still impossible to answer these questions. Nevertheless, by the time he climbed out of the water, something was different. His mind had changed; his thoughts had settled. Starting tomorrow, he would focus on spinning the X-wing's thruster—even if it meant scrapping everything and starting over.

- V -

LAYER FOUR: THINKING ABOUT THINKING

Sometimes I like to think about the laws of learning as a map of a mysterious sea. At the dawn of the twentieth century, E. L. Thorndike described some of its most significant geographical features, and ever since, generations of cartographers have attempted to add detail and, as needed, correct his mistakes. There's just one problem: For those who have to actually ply these treacherous waters—learners and their teachers—it's not clear which faction of modern cartographers to trust. On one hand, you have inside-out maps created using mechanistic models: "Thanks to what we know about coastal geology, there should be dangerous submerged rocks here." And on the other hand, you have outside-in maps, based more on practical observations: "Ships that head out in this particular direction often never return, so that must be where the dangerous rocks lie."

Both strategies have steered learners astray in the past. It was the outside-in impulse, for instance, that led many school districts to experiment with Whole Language. The story remains a cautionary example of how ignoring the nitty-gritty of brain function can lead to pedagogical ruin.

But inside-out thinking, too, can lead to missteps. In 1983, the Harvard psychologist Howard Gardner, observing that brain damage due to a stroke or other trauma can interfere with, say, reading, but leave numeracy and other abilities untouched, began arguing that the brain has not one overarching intelligence but rather many separate, domain-specific intelligences. This multiple intelligences theory remains, to this day, a strong, well-supported indictment against the very idea of general intelligence. As Gardner has put it, IQ tests may put too much weight on certain abilities simply because they're easy to measure: a case "of the man looking for his dropped car keys underneath the street lamp because that is where the light is." His alternative theory, inspired by mechanistic ideas of how the brain works, served as an inside-out update on older inside-out thinking.

To Gardner's dismay, however, a group of true believers soon took his theory too far, trotting it out as evidence of what became known as multiple "learning styles." This noxious idea, which holds that most students require specialized education media depending on their supposed brain makeup, lingers zombie-like in education culture despite a wealth of evidence against it. I still routinely encounter students claiming to be "auditory" or "visual" or "kinesthetic" learners. Perhaps they do possess certain relative aptitudes—clinical psychologists, after all, routinely invoke granular intelligence assessments in their diagnosis of specific learning disorders—but that doesn't, and shouldn't, change my classroom approach. Every large-scale review on the subject has shown that matching the delivery of educational content with students' preferred learning styles fails to produce any measurable beneficial effects, and may even hurt long-term retention of knowledge.

Also to be laid at the feet of neuro-minded, inside-out reformers is the myth of "right brain" and "left brain" dominance, which holds that certain students are more creative or logical depending on which hemisphere prevails in their thought, in the same way that most people are right- or left-handed. This notion can be traced back to the nineteenth century, but received a major boost in the 1960s thanks to the crude-in-retrospect procedure of surgically separating patients' hemispheres in hopes of curing epilepsy. This practice revealed that verbal language couldn't be summoned without input from the left hemisphere, but in the years since, fMRI research has shown that almost all cognitive tasks invoke both hemispheres to some degree. In any case, if someone appears to be more creative or more logical, there's no reason to assume that their proclivity arises at the level of gross brain structure and not at a higher, psychological level. The concept of right-versus-left-brain dominance may still be marginally useful in some cases, but only as a metaphor.

There exist other neuromyths, such as the false notion that we only use 10 percent of our brains. (Modern-day inside-out thinkers bear no blame for this one; it can be traced all the way back to William James.) In fact, even in the Whole Language debacle, for which outside-in logic bears most of the culpability, inside-out partisans also had a small role to play.

In the late 1980s and 1990s, teachers entered into a tempestuous romance with a practice then known as "neuroeducation." The idea was built around new research into childhood brain development. Kids' synaptic connections, it turned out, increase in raw number up until about age ten, whereupon the brain commences a years-long synaptic "pruning" routine via a mechanism that is essentially the opposite of LTP. This process molds the brain into its adult configuration by one's early twenties, when it delivers the finishing touches to the prefrontal cortex. This region is highly involved in executive function—not just working memory but also attention, emotional regulation, and staying on task. I've always thought that its laggardly development must explain some of my behavior at university.

If the years between birth and age ten were prime time for synaptic development, the neuroeducation reasoning went, then it was imperative to flood youngsters' sensory worlds with data-rich stimuli before it was too late. "With the right input at the right time," a *Newsweek* writer opined in 1996, "almost anything is possible," but "if you miss the window you're playing with a handicap."

There remains some truth to the idea that certain types of learning are easier to pull off during certain critical periods in development, verbal language learning chief among them. But in the mid-1990s, so magical did some consider the period of early synaptic growth that multiple U.S. states enacted legislation devoted to such sensory-enriching schemes as the playing of classical music within earshot of young children, based on a since-debunked idea that mere exposure to Bach and Mozart would lead to improved spatial reasoning capabilities down the line. This practice was as harmless as any other placebo, in the sense that it only caused problems when it got in the way of better-substantiated methods. Which, indeed, was exactly what happened in the case of Whole Language. With the magic period of synaptic creation backing the claim that kids could somehow make sense of words on the page if only they saw enough of them, lawmakers and school boards signed on. The effect, in some states, was a generation of students that lagged far behind their counterparts in reading.

By the mid-1990s, however—even before the problems with Whole Language were widely acknowledged—neuroeducation skeptics were arguing that educators should think twice before making a leap straight from brain science to educational practice. "Cognitive psychology is a much better bet" than neuroscience, wrote the Washington University cognitive psychologist John T. Bruer, in the most important document of what came to be known as the "neuroskepticism" movement. "Currently, the span between brain and learning cannot support much of a load." Instead, he suggested, for brain research to contribute to successful educational practices in the future, it would be best if it were to follow an "indirect, two-bridge route," first pausing for verification at the level of cognitive psychology before continuing along to the classroom.

That's perhaps a viable pathway for some inside-out thinking to find its way into useful educational practice. But then again, it still doesn't account for the fact that even the most fair-minded educational psychologists can cherry-pick supporting neuroscience research when it suits their purposes, which can lead to fiascos like Whole Language.

Happily, there is another way. We can strategically integrate inside-out and outside-in approaches by looking back through the history of the cognitive revolution for psychological theories that both update Thorndike's laws of learning in significant ways while also remaining plausible according to the findings constantly coming out of neuroscience. One such theory of learning does, in fact, stand out. Its key insight is as surprising as it is simple: The secret to better learning may be better forgetting.

THE POWER OF FORGETTING

Over the decades, the vast majority of updates made to Thorndike's laws have had to do with his law of effect, which describes how memories are stored in a web of associations. His law of exercise, which has to do with forgetting, has received far less attention, perhaps because most people, Thorndike included, have long considered forgetting to be nothing more than the loss of memory.

Thorndike's thesis was simple: Memory decays as a function of time elapsed since it was last accessed. That the reality was far more complicated soon became apparent, however. In 1932, the Missouri psychologist John Alexander McGeoch asked study subjects to memorize random pairs of words: *Please associate the words "cat" and "peanut," so that whenever I say "cat," you say "peanut."* The simple task became measurably more difficult when, in the interim between learning and recitation, McGeoch presented his subjects with confounding word pairs: *cat* and *glasses*, or *cat* and *typewriter*—which turned out to make remembering the original pairs far more difficult than Thorndike's time-based model of decay would have predicted.

McGeoch's explanation changed the very meaning of forgetting. Study subjects, he argued, failed to remember *cat-peanut* not because that memory was erased outright, but rather because other, competing memories got in the way. Soon, interference, not disuse, became the leading explanation for forgetting, and this came with profound new implications for what constituted a memory.

Suppose a memory is an old house, deep in the woods. If interference, not disuse, had made it inaccessible—say, vandals blazed decoy trails leading away from the main access trail—that didn't mean the house disappeared. It was still back there, just inaccessible at the moment. Many researchers (including even B. F. Skinner) began to consider the state of such access trails and the state of the memory itself—the house—as two wholly separate questions. Different researchers over the years gave them different names, but the ones that stuck are the terms "retrieval strength" and "storage strength," coined by the cognitive psychologists Robert and Elizabeth Bjork. As Robert succinctly puts it: "Storage strength is how well learned something is; retrieval strength is how accessible (or retrievable) something is."

When an item has low storage and low retrieval strength—say, something you can't remember from a long-ago class to which you never gave your full attention—it's not much of a memory at all. Something recently learned and of temporary importance, meanwhile—say, Mikael Lundqvist's office number—has high retrieval strength, but low storage strength, which means it is temporarily accessible but readily forgotten. There are also items of memory with high storage strength but low retrieval strength: The name of that melody running through your head, or perhaps your childhood best friend's phone number, which you can no longer recall. You might not be able to remember it in the moment, but given the right cue, the memory would come flooding back.

Finally, a memory with both high storage and retrieval strength would be something you know well and can easily retrieve at will: the name of the first American president, say, or your birthday. Or perhaps the name of someone who changed your life with an act of unexpected generosity.

A physics-turned-math major from the University of Minnesota with a John Wayne vocal rhythm and abiding love for the game of golf, Robert Bjork joined the young field of mathematical psychology in graduate school. He had put in one graduate year at Minnesota when his advisor, David LaBerge, told him that the field's major figure, William Estes, was moving to Stanford. Bjork was an impressive student who held great promise for the Minnesota program, and yet LaBerge urged him to move to California anyway. "I mean, it was so much in my interest and so much against his own interest. It's just kind of amazing," Bjork later recalled.

Estes, meanwhile, had been a disciple of B. F. Skinner's prior to the outbreak of World War II, when Estes was sent to serve in the Pacific theater. During the war, according to the story Bjork heard, family members back home had sent Estes books to read during his downtime, but there were strict weight limits on what they could send, and he devoured the books far faster than they came in. Frustrated, he wrote home and asked for a book on mathematics. "He wanted to learn more anyway," Bjork said, plus there was an added benefit: A serious math book would take a long time to read. Estes returned at the end of the war with new mathematics knowledge in tow, and went on to help found the field of mathematical psychology and the journal of the same name.

Mathematical psychologists attempted to model simplified learning brains, a line of work that formed a front of its own in the cognitive revolution, and which ran in parallel with the sort of computational research Seymour Papert and others were undertaking at MIT in the 1950s and 1960s. By 1966, Bjork, now under Estes's wing, found himself feeding punch cards into one of Stanford's handful of mainframe computers, trying his best to make a mathematical model of learning, known as a Markov chain, work. A Markov chain is a probabilistic sequence of states—a topic that would later come up in my own research at MIT—and mathematical psychologists were pioneering their use in simulating how a learning brain might function. Bjork had tasked his simplified brain with remembering pairs of items, just like in McGeoch's study thirty-odd years prior. His model stipulated the existence of a short-term and a permanent memory state, but there was a problem: information in the former didn't seem to want to pass into the latter. "There was no transition to this sort of permanently learned state," he said, which "was kind of shocking at the time." What was more surprising, however, was Bjork's solution: He let the short-term memory units engage in random acts of forgetting. With that, the model began obediently chugging along, the to-be-remembered information sliding perfectly into long-term storage. The outcome sent Bjork on a career-long journey—first, to the University of Michigan, then to his eventual academic home at UCLA—to figure out why forgetting was so important, and whether what appeared to be true in classroom-sized computers could be said for classroom-dwelling humans.

—

And so it was at the UCLA Faculty Club, under a potted palm, that Robert and Elizabeth Bjork, Robert's longtime collaborator and spouse, graciously agreed to sit for an interview. The Faculty Club is a single-level oasis of a building, situated on a bluff, overlooking a surface street chockablock with overstimulated motorists. Although the campus's famous inverted fountain wasn't running, Los Angeles was experiencing a welcome reprieve from the long drought that had plagued it for years, as well as a spate of wildfires that had overcast the city with smoke the month prior and would return with a vengeance a month later. It was as though someone had hit the pause button on the unfolding environmental calamities, just long enough for a conversation about how to better leverage our species' brainpower.

Robert and Elizabeth sat down with cappuccinos.

If, in the years following Robert's turn to forgetting, it was obvious to psychologists that there was a difference between memories' storage and retrieval strengths, it wasn't clear what exactly to do with that information. Together, however, the Bjorks would go on to put forward an intriguing theory that didn't just tie the two together, but would position forgetting as fundamental to how the learning brain works.

"What was really different, I think, and new with us, was this idea, not that you needed these two strengths," said Elizabeth, "but that they interacted with one another."

They had had their suspicions about the importance of forgetting virtually from the start. After all, William James had suggested back in 1892 that "if we remembered everything, we should on most occasions be as ill off as if we remembered nothing." The Bjorks had been inclined to agree: Without forgetting, Robert predicted in 1972, we would "degenerate" to a "state of total confusion."

In an attempt to dig deeper, Robert—Elizabeth would formally join later—began to explore different ways to make new memories irretrievable. In a stream of studies running the length of the 1970s, he and his research partners used a variety of tricks to lead their study subjects to the shores of forgetfulness. The first of these involved interleaving to-be-remembered information with distractions, in the tradition of McGeoch. Robert also tried physically moving his subjects—one room for learning, another for assessment—the idea being that cues from one's senses contribute to which memories you can retrieve at any given moment. And he explored spacing out study sessions temporally, which led to forgetting in the short term, as anyone who habitually crams for exams could tell you, even if it's beneficial for long-term retention. All of these tricks provoked measurable forgetting relative to control groups.

He even tried to induce forgetting by simply instructing his study subjects not to remember certain items. After a group memorized a long list of information, Robert told them to forget it: that list had just been for practice, and the real to-be-remembered list was upcoming. After they had memorized the second list, Robert tested them on both anyway. Somehow, his instructions had gotten through: compared to a control group, their recall of the first list was impaired, and the second list enhanced.

Far more interesting than all the various ways Robert induced forgetfulness in this multiyear series of studies, however, was what happened afterward, when he gave his subjects cues to help them remember what they'd lost. Thus refreshed, the memories didn't just return, but came roaring back as though they'd been given a new lease on life and wanted to make the most of it. These forgotten-then-rejuvenated memories appeared to be both more readily accessible and stickier than memories that had never been lost in the first place.

For a common example of the power of forgetting-then-relearning, Robert suggested thinking of an all-too-common social challenge: "You're at some meeting or party or something where you really want to try to remember the names of the people you're meeting," he said. When confronted with a new name to remember, "One thing people do when they're really concerned is kind of repeat it over and over to themselves. Not out loud, of course." The tactic helps, but only for a short time: "That won't do anything as far as creating long-term learning," he said. And so later in the night, you find yourself faced with competing possibilities: Was he James, John, or Jake? Suddenly, an hour later, the name comes back: Jim, of course. Having forgotten a name, says Robert, then, "at some time later, looking across the room and retrieving what that person's name is—that can be a really powerful event in terms of your ability to recall that name later that evening or the next day."

Relying on just the power of forgetting and retrieval alone, students and teachers alike can glean impressive ways to improve their study habits and instructional tactics. According to Robert's early results, for instance, the long-lasting mnemonic benefits of spaced-out study sessions could be explained not only by synaptic strengthening at the cell level, but also at a level higher up the cognitive high-rise, where spacing boosts retention by making forgetting and re-remembering possible. "There have been arguments off and on about the mechanisms" involved in spacing, said Robert, "and sometimes we think that all the explanations may be correct. They're not mutually exclusive." Mix in the confounding effect of interleaving, as I've noted—alternating one spaced-out study session with another—and you have the potential for deep, robust memories, provided, of course, you eventually retrieve what you forget during these exertions.

But the Bjorks were just getting started. Of particular interest was Robert's finding regarding directed forgetting, where he had simply told his students not to remember something, and they had miraculously complied. What mechanism, he wondered, could possibly produce an effect like that?

—

During the 1970s and most of the 1980s, Elizabeth had to content herself with a background role in this line of research.

"There used to be nepotism laws," explained Robert.

"That was part of it," she agreed.

"For many years those laws penalized women. If a couple went somewhere, the man got the job," he said. "When we came to the universities here, I think we were once told we were the second or third couple in the whole—"

"System," she said.

"UC system," said Robert. "But then there was this period where there weren't the nepotism laws anymore, but there were informal things like, 'you shouldn't publish together,' or 'you needed to work in different domains.' So even though Elizabeth's background was learning/memory, she had an era of working on—"

"I turned myself into a visual cognition person for a while," she said.

"And did some things on children," he said.

"And I also did some developmental stuff," she said. "Although we'd been working and talking the whole time."

"The whole time, working and talking," he agreed.

Including during the period leading up to a 1978 study Robert coauthored, which was aimed at explaining the "mystifying ability on the part of subjects," as the authors put it, to forget items seemingly on cue. The effort was only marginally successful, but the finely tuned methods they employed happened to reveal something unexpected. The *act* of retrieving items from memory didn't just boost their later retrievability, but also actively depressed recall of competing items.

With the unexpected finding that the act of retrieval can modify memory *beyond* the item being recalled, the Bjorks had their *mise en place:* ingredients laid out, tools at hand. And, at long last, both *chefs dans la cuisine*, ready to take on the master of learning, E. L. Thorndike.

Slowly, universities began to come around on the idea that married couples could work together, to the point where today, hiring a couple at the same time has become a recruiting tactic for many schools. And finally, after years of collaboration if not coauthorship, the Bjorks produced a paper that, upon its release in 1992, resounded like a hymnal dropped in church.

It was "the first thing we published together," Robert said, "twenty-three years after we were married." It was also perhaps the single most significant update ever applied to Thorndike's original map of how learning works.

THE NEW THEORY OF DISUSE

If forgetting were purely a function of mnemonic decay, as Thorndike posited in his theory of disuse, then it was purely harmful: the sort of eternal enemy, like hunger or cold or death itself, against which we animals have evolved to valiantly struggle. Remembering everything we encounter, the thinking went, must be expensive in terms of either bodily energy or data storage, and so at some point, forgetting becomes inevitable.

In their 1992 publication, the Bjorks flipped that idea on its head. What if, they mused, long-term memory capacity is, for all practical purposes, limitless? After all, up and down the cognitive high-rise, researchers had come to adopt a picture of memory that looked less like a finite container to be filled and more like a growing tree, with every new branch only adding to the total available storage capacity. In such a model, might forgetting function less as an unavoidable yin to memory's yang, and more like a *tool*, evolved over the eons, for pruning the overgrown Fragonard scene that is our tree of knowledge?

Making such a tool possible was the interplay the Bjorks had observed between retrieval strength and storage strength. Moments after you first meet someone at a party, it's very easy to repeat their name to yourself, since retrieval strength is high. Meanwhile, in the moment you register that name, its storage strength climbs from zero to a low level. But repeating the name does little to increase that storage strength. The reason why, according to the Bjorks, is that there are diminishing returns to revisiting a memory while retrieval strength is high.

"The higher it is in retrieval strength, when I produce it or study it, the lower the increase in storage strength," Robert explained.

Only once the memory's retrieval strength drops as a result of interference, whether due to the passage of time or encroaching sources of confusion, do you get another crack at increasing storage strength in a more significant way—such as at our hypothetical party, when you dredged up the once-forgotten name after a time lapse (Do you remember what it was?), and presumably attached new associations to it as a result.

If you're trying to remember something, said Robert, "storage strength, just vaguely, would be how linked up is that thing with related things in your memory. So, if it's some event, or a friend or teacher's name from your past, maybe all the way back to high school and stuff, it will be linked up with lots of things. Images of the school, episodes you did together, semantic things."

The sheer ballast of all those connections helps explain why it's almost always easier to relearn information a second or third time around. It also affects retrieval strength. "That's another thing in our theory, that the rate at which you lose retrieval strength depends on what the storage strength is," said Robert. When storage strength is very high, retrieval strength fades so slowly that it might never go away—which is why a healthy person will always be able to summon up the name of their mother, or George Washington. And even when a memory's retrieval strength does fade, the storage strength remains, ready to rekindle that retrieval strength at a moment's notice. "As you move through your life and you change contacts, and meet other people and so on, the name of the best friend, or the best teacher you had, or whatever, won't be recallable, but it will be recognizable," said Robert. "And if something—going back to a class reunion or something—provides enough cues that you start to recall and use it, there will be a huge increase in retrieval strength that will become very available for quite a long time."

Put another way, if the experience of forgetting is actually the loss of retrieval strength due to interference from other thoughts and memories, then forgetting dials those interfering associations down, close to zero. Retrieval reblazes the true path. The other competing paths, meanwhile, fade relative to the correct path, because the correct path is interfering with *them*.

But when storage strength is low—when that house in the woods is more like a lean-to—retrieval strength fades fast. This is why you can't recall the name of the person at the party even minutes after meeting him, no matter how many times you recite "Jim" to yourself.

To pull an actionable takeaway from the Bjorks' theory, then—one that overlaps almost completely with findings from the synaptic level of the cognitive high-rise—whenever a memory feels easy to retrieve, doing so will not add much to its storage strength. But *effortful* retrieval, made difficult by the strangling vines of competing associations, adds to storage strength, and in turn preserves retrieval strength for the long haul.

In terms of updating Thorndike's map of the learning brain, this notion of effortful retrieval soon took on great importance. By the middle of the twentieth century, Thorndike's model of decay had received strong metaphorical support in the form of electronic computer storage. Perhaps, many reasonably decided, we forget things simply because we run out of space, like a hard drive filled with photos. As the renowned ichthyologist David Starr Jordan once complained upon becoming president of Stanford University, "Every time I learned the name of a student, I forgot the name of a fish."

In the Bjorks' theory, however—which, in a direct shot at Thorndike's theory of disuse, they named the *new theory of disuse*—there was no limit to the brain's storage capacity. There were, however, theoretical limits to the amount of information that could conceivably be *retrieved* at any given moment.

"As we make some items in memory more and more accessible, according to our theory of disuse, there is less and less remaining retrieval capacity for other items. This viewpoint, then, may exonerate the ichthyologist David Starr Jordan," the Bjorks wrote in their seminal paper outlining their theory. "He is often cited uncharitably as someone who had a fallacious idea of the capacity of human memory. Given the limit on retrieval capacity assumed in our theory, however, an ichthyologist suddenly spending considerable time learning and retrieving the names of a large number of students could well lose access to the names of certain fish."

If retrieval strength were indeed a relative measure throughout the brain, then it was also relative on a more local scale. Each memory representation would have a variety of relatively strong and relatively weak paths leading in. And each "cue"—a sensory or cognitive impression that triggers a train of thought—would have a variety of relatively strong and relatively weak paths leading out. In the Bjorks' model, then, effortful acts of retrieval (always in the wake of forgetfulness) apply a Rototiller to your cognitive landscape, prioritizing paths from cues to helpful memory representations while weakening paths that might lead you to confusion—*cleaning away*, in effect, superfluous and faulty associations.

"Overall, it's kind of adaptive the way it works," said Robert. Forgetting "frustrates people because obviously we're all familiar with not being able to recall things we want to recall. But on some broad statistical basis, most of what we're able to recall is things associated with the current contextual cues—things that are more relevant to current tasks and this phase of our lives." Forgetting, therefore, ensures access to the most important information for the present, even as less immediately useful information is stored for a rainy day—out of sight, perhaps, but never out of mind.

EFFORT GOES TO SCHOOL

As tidings of the new theory of disuse spread, educators looked for ways to put the theory to work: to trace the latest updates from the scientists' maps onto the more practical charts currently in classroom use. It soon became clear that the theory could be applied almost anywhere learning was expected to occur.

Robert, for instance, began to direct a sliver of his attention toward effortful retrieval in the context of motor-skill learning, which gave him the opportunity to revisit one of his first loves: golf. In a classic study from 1978, researchers had assessed the ability of two groups of children to throw a beanbag into a target three feet away; one group practiced only at that distance, while the second group practiced from either two feet or four feet away, alternating. At the end of the study, this second group outperformed the first group at three feet away, despite never having practiced at this distance. This result came as strong evidence of the power of variation, but there had been no mechanistic explanation forthcoming at the time. Now, well aware of the plausible mechanism of effortful retrieval, Robert began noticing suboptimal training patterns everywhere—particularly at the driving range, where people typically train on one club ad nauseam, switching to a new club only after a long period of massed practice.

"People do everything wrong on the driving range," he said. "They hit a good one, they rake a ball" over, and repeat, again and again. "That's just short-term motor-memory change. They don't interleave, they don't space." The approach can lead to noticeable improvement *within* a given training session, but by the early 1990s, a growing list of studies was showing that such blocked practice was suboptimal for the long-term retention of skills, as well as transfer of skills among different tasks. One group of researchers, for instance, had badminton players practice three different types of serves in either massed or interleaved sequence. The former made faster progress during the instruction period, but following a delay, the latter group was not only more accurate, but also better able to serve from the opposite side of the court—a position where neither group had practiced.

One motor-learning idea from the mid-1980s that gained a new mechanistic explanation in the Bjorks' theory was the "reloading hypothesis," which holds that massing one's golf practice is harmful because it allows you to economize retrieval of the motor program responsible for your swing. Golfers often "load" the program only once, even if they hammer out ten or twenty repetitions in a row. "They don't ever reload the motor program," Robert said. But switching to a new club or picking out a new target with every repetition requires you to reload the relevant motor program each time. The effect is more effortful retrieval, and thus stickier and more accessible skill memories. Robert, who has shot his age ninety-six times since he turned seventy-two, has spread the gospel of skill interleaving throughout the hallowed halls of golf, including at the annual meeting of the PGA.

Back in the classroom, effortful retrieval has helped explain why interleaving tends to produce better results than spacing alone. A more recent technique, of which Elizabeth Bjork has become a particularly outspoken proponent, takes effortful retrieval even further. If retrieval can strip away faulty associations and rejuvenate memories, why wait until exam time to retrieve the answers to questions? Recently, Elizabeth has advocated for the practice of *pretesting*, which is exactly what it sounds like: taking a practice test prior to the real deal. Like so many threads relevant to the new theory of disuse, research into pretesting can be traced back to the 1970s, when the first of many studies came back with results showing that pretesting leads to marked improvement on final exams. More recently, Elizabeth coauthored an intriguing article showing that multiple-choice pretests can even improve retrievable knowledge related to the *incorrect* answers. For instance, answering a question about the oldest geyser in Yellowstone National Park (Castle Geyser) also activates the potentially useful knowledge that Old Faithful must be younger, and does so more effectively than would passively reading that fact.

Despite how well the technique works, however, students often meet pretests with *pro*tests. They take time, and more importantly, the effortful retrieval they demand takes energy. This issue raised one question that had been bothering the Bjorks since before they published the new theory of disuse. Why, precisely, does effortful retrieval have to *feel* so difficult? After all, there are plenty of expensive activities—in terms of both calories burned and cognitive resources utilized—that the brain rewards, ranging from curiosity to sexual behavior. What was it about effective learning strategies that made them seem so difficult?

DESIRABLE DIFFICULTIES

Answers to that question had already started to cross Robert Bjork's desk as early as the mid-eighties. In 1985, Robert had been asked to join a committee of the National Research Council, underwritten in great part by the United States military, tasked with investigating any and all techniques that could conceivably produce better soldiers, faster. The Committee on Techniques for the Enhancement of Human Performance, as it was known, left no learning-related psychological theory unexamined, no matter how outré. Soon enough, to the dismay of science fiction fans everywhere, it conclusively reported "no evidence of the existence or usefulness of elements of parapsychology, including ESP, telepathy or thought projection, and mind-over-matter psychokinesis," as the *New York Times* detailed in 1987. The committee was meant to be a one-off effort, but new research questions popped up as quickly as it could address them, and its work continued well into the mid-1990s, when Robert assumed the role of committee chair.

During this time, amid such distractions as ESP and parapsychology, a handful of promising, serious research threads began to appear. Some of these, such as learning during sleep via audio recordings, led nowhere. The idea of metacognition, however—that is, how we think about our own thinking and learning—proved so interesting, and fit so snugly with the new theory of disuse, that it went on to influence not only the rest of the Bjorks' respective careers, but, ultimately, maps of applied learning trusted by educators in the field.

—

In the history of learning research, the idea to study metacognition came surprisingly late. Facts like the power of effortful retrieval "have been known one way or another for a very long time, but it's almost like they can get forgotten sometimes in the domain of education," said Robert. "But the field of metacognition is much younger. It took, I think, some of us quite a long time to see certain relationships."

In theory, the subject should have fit right into the wheelhouse of researchers like Thorndike, who led psychology away from the navel-gazing methods of "introspection," as it was known, and toward quantitative techniques that would yield hard data. Although both Thorndike and Dewey variously flicked at it, metacognition proved too slippery for quantitative analysis. There was no obvious reason to doubt one's own subjective sense of having learned something or not, and so psychologists rarely bothered to ask how those sorts of impressions could deviate from reality.

In the 1960s, however, that began to change. In the first modern metacognition study, the psychologist Joseph Hart compared people's *feeling-of-knowing*, as he dubbed it, against known facts. He worked from the assumption that we use some sort of static metric, like an engine oil dipstick, to judge our own state of knowledge about a topic. The follow-up research that caught Robert Bjork's eye, however, suggested a different automotive analogy: a speedometer. In the same sense that a car's speedometer infers ground speed from engine rotation, we infer knowledge about the state of our memory based on the speed and effort involved in encoding and retrieval.

This speedometer can, however, be fooled, which is especially distressing in the light of the many studies showing that students tend to study a topic until they reach what feels like an acceptable level of knowledge, but no further—a recipe for disaster if your judgment of your own knowledge is off. And indeed, almost every aspect of what we know about what we know, and our ability to gain more of it, is fraught with potential inaccuracy.

For starters, psychologists have identified three major sources of metacognitive bias regarding our ability to add to our own memory. When information is laid out in front of us, *hindsight bias* causes us to treat it as though we knew it all along—which becomes a problem come exam time, when it turns out that what you thought you knew is stored in your textbook, not your memory. *Foresight bias*, meanwhile, is an artifact of how our memory is structured. A given cue may reliably call up an answer, and we believe (at our own peril) that we will be able to summon it again in the absence of that cue. For instance, plenty of students might be able to answer a true-or-false question concerning whether hemoglobin accepts oxygen in the blood, but won't be able to name the molecule when asked point-blank. Finally, many of us demonstrate what Robert Bjork and coauthors have dubbed a *stability bias:* the false sense that what's accessible or inaccessible in our memories will remain that way over time. In fact, the information you can access shortly after instruction is a poor guide to what you'll be able to access at a later date. Stability bias translates readily into both overconfidence (I already know everything I need to know, so I don't need to study) and defeatism (I don't get this topic and I never will, so why bother studying).

In addition to the general distortion field cast by these common, overarching biases, other factors can give us a wrong impression of what we know, including how we've retrieved and encoded a given set of facts. If you can call a fact to mind quickly and easily, for instance, you might assume that it's correct—but it might just be a well-learned falsehood. Meanwhile, facts that feel easy to learn in the moment don't always remain easy to retrieve. That's especially true if they felt easy to learn due to the format in which they were presented. Quite the contrary, in fact: In a remarkable series of studies, learners described words that were delivered in larger type, and at louder audio volume, as more memorable than quieter and smaller words—an impression roundly disproved by subsequent memory tests.

Robert Bjork, sensing a pattern nearly two decades before many of the studies referenced above came out, realized that metacognition was more than just a neglected facet of cognitive psychology. In fact, it fit perfectly with his and Elizabeth's new theory of disuse. Perhaps, he reasoned, metacognition research could combine, Voltron-like, with the new theory of disuse, and do what neither could do alone: break into widespread educational practice. On one side, you had learning techniques that were untapped because there had been no room for them in Thorndike's original rules of learning; on the other, you had techniques that were going unutilized because students mistakenly found them ineffective, and thought that they made learning harder.

Upon the techniques that ticked both boxes, Robert bestowed the name "desirable difficulties." Spacing instead of massing certainly made the list. It could neutralize the false sense of security given by recently learned facts; plus, by giving retrieval strength time to subside, it gave learners a chance to shore up storage strength; *plus*, as an additional bonus, there were still all those benefits at the synapse level that neuroscientists had been talking about since the 1970s.

Interleaving created even more desirable difficulty: it accomplished everything spacing did, only with more forgetting and therefore better improvements to storage strength, as well as more transfer between disciplines. Pretesting fit the bill as well: It demanded effortful retrieval and came with the added benefit that pretests didn't lie about the state of one's knowledge, unlike one's own feeling-of-knowing. In Robert's book, even utilizing an unexpected or slightly difficult-to-read font qualified as a desirable difficulty.

In the mid-1990s, the Bjorks sent the idea of desirable difficulties out into the world through the normal academic channels, and then they waited. A handful of interested parties latched on, but the bulk of teachers and students remained oblivious. Then, in 2014, a pair of the Bjorks' occasional research collaborators, with the help of a professional writer, published an excellent book filled with insights from cognitive scientists, including the Bjorks, titled *Make It Stick*.

One reader of the new book was Louis Schulze, the new head of academic support at Florida International University (FIU) School of Law, a few blocks and a handful of bridge-linked islands from Miami Beach. For years, the school's bar-exam passage rate had bounced around the middle tier of Florida law schools: number seven out of the state's eleven law schools in 2012, up to number three by 2014, then down to nine as of February 2015. Schulze was charged with turning around the school's lackluster rates.

Traditionally, law schools looking to make big changes in bar-exam passage without changing the composition of incoming classes had sought to do so via what Schulze called "silver bullets": pouring resources into a class (usually civil procedures) or set of classes that seem to wield an outsize effect. A data-oriented colleague at FIU, Raul Ruiz, who arrived around the same time as Schulze, ran the numbers, however, and showed that this approach had never worked at FIU. It was time to try something new, and Schulze decided to read the studies described in *Make It Stick*—including the Bjorks'.

That year, he began a program, running in parallel to FIU's traditional law curriculum, designed to teach students how to absorb the often-absurd amounts of information required of law students. Today, the program unfolds over all six semesters. First, Schulze offers a voluntary, zero-credit course, designed to teach students such tactics as self-pretesting, as well as an optimized approach to outlining course content (a common law-school study method) that he has developed. "They have absolutely no compulsion to be there whatsoever," he said. However, in the second, third, and fifth semesters, if you're in the bottom 20 percent of the class, continued instruction with Schulze on learning to learn becomes mandatory. And finally, "the sixth semester course is the big one," he said. It's optional, but "90 to 99 percent of the class signs up for it now." At this point, he said, "they've seen the success numbers."

Schulze's program is designed not to add to the legal information delivered in students' other courses, but to complement it using tools like desirable difficulties. "We're not re-teaching law," said Schulze. "As matter of fact, we've got a 'flipped classroom' model where they have to teach *themselves* the law outside of class, and then when they come into class it's the testing effect, it's metacognition, it's spaced repetition."

For some students, the program is a godsend. The rest of the classes at FIU are still structured like normal courses, with high-stakes finals that would, at least in theory, reward short-term study techniques like cramming. But there's a catch: In law school, there's too much material to fit into even a several-day cram session. "It's not like an undergrad course where you can maybe memorize ten pages worth of material, walk in and then vomit that material on your essay and be good to go," Schulze said. "If you outline a Constitutional Law class thoroughly, it's going to be 120 pages." To convince his students to space out their study sessions, he pulls out time-honored evidence: a version of Hermann Ebbinghaus's forgetting curve, applied to legal knowledge. "Basically, you take a class on intent." Two days later, your ability to retrieve that information is "down to 30 percent. And then what law students do is, they don't touch that material again until ten days before the exam. And so it flatlines throughout the entire semester at 30 percent, and then when they try to cram at the end, it only comes back up to, like, 45 percent. And so I tell them if you use spaced repetition, instead you're going to walk into an exam with 80 percent of knowledge, but your colleagues are walking in with 45 percent knowledge."

In 2015, just two years into Schulze's program, FIU rocketed from ninth place on the state's bar passage list to first place. As of 2019, it's maintained that position in seven out of nine of the state's semiannual exams administered thereafter, never dropping below second place. The school's "ultimate" pass rate—the percentage of students who pass the bar exam within two years of graduation—is now ranked within the top fifteen in the United States.

"We're teaching them how to teach themselves better," Schulze said.

THE ANTI-WINNOWER

Perhaps the most inspiring aspect of FIU's Academic Excellence program, as it's known, is what it does for students who might not otherwise manage to fight their way into the legal profession. The key, says Schulze, is the second- and third-semester courses, mandatory for students in the bottom 20 percent of their class of roughly 140 students. "A good chunk of those people, probably ten of them, are failing out. They're at a GPA below a 2.0. And if they don't get it up above a 2.0 by the end of the semester, they're academically dismissed," he said. "So at that point I've got some folks who are just like, 'All right, I'm freaking out. I'll do what you're telling me.'"

They don't all adopt Schulze's methods, and they don't all succeed.

But enough of them do. Enough to make a difference.

"That's the group who, if you see why we're now in first place, that's the group that really moved the needle for us, because they went from a pass rate of something like low 60s, high 50s, to now they pass it in the high 70s, low 80s," he said. "I know anecdotal evidence isn't proof of anything, but one of our students last year, who was, like, number three from the bottom of the cohort, he just totally jumped into spaced repetition." Over the course of his three-year law school career, he went from complaining about what he considered to be a sieve-like memory to raving about spacing. Finally, as he walked into the bar exam, he'd been pretesting at about 70 percent, a decent margin above the 63 percent needed to pass. "So he was very comfortable going into the exam," Schulze said.

Results like these are always uplifting—especially when they come to students as deserving as Schulze's. "We're very fortunate to have students whose background is such that they work really hard. We have a lot of first-generation Americans," he explained. "We don't have that problem that some other schools have where the students feel entitled to pass the bar exam. We don't have that. Our students scrap for it. And so I think that, while some of our students may have come to law school from less privileged backgrounds, they've got the intelligence, they've got the aptitude, they've got the hard work," he said. "When we just showed them, 'here are the best ways to teach themselves,' it just unlocked their natural abilities."

On the subject of FIU Law, said Robert Bjork, "Overall and anecdotally, it has been remarkable when students have changed their own routines to incorporate some of these desirable difficulties."

Perhaps what's even more remarkable than the FIU results is the simple fact that the Bjorks have shown that change is possible. Disagreement about what constitutes improvements to education may still linger everywhere in educational psychology, and even in neuroscience. But at the very least, the Bjorks have shown that updating the maps—both the scientific charts of the learning brain, as well as practical maps for how to teach and learn—can have immediate beneficial consequences.

The Bjorks' research even carries critiques for the eternally warring outside-in and inside-out camps. For those in the former, who might assume "learning to learn" comes naturally in a project-based curriculum, the Bjorks have strong words. Said Elizabeth: "So much of our research has sort of shown emphatically that," in deciding how to best take in information, "if you just go with your gut feelings, your instincts, what you think sounds good, what should work—most of the time it's incorrect."

Meanwhile, the inside-out camp wades into equally dangerous territory in its love affair with worked examples. There are no doubt situations where "worked examples are better overall," said Robert, "simply because, perhaps, the learner in a given situation isn't succeeding enough," and they need help getting over a problem-solving hump. However, "I think just one key—and this goes across sports and everything, I think—is just to get the learner to produce something, one way or another." Worked examples short-circuit effortful retrieval, he explained, which remains the key to making memories available in the long term. "There's almost never going to be anything quite as powerful as that. No way."

As inspiring as the story of desirable difficulties may be, however, it comes with a sobering dark side. Take Schulze's success: a wonderful story, except for the fact that it shows that in *most* mid-tier law schools, students like his must be flunking out en masse—a failure of education, a squandering of human potential, that we should now recognize as entirely preventable. Robert Bjork and a pair of coauthors made a similar point in a 2013 article on metacognition. "There is, in our view, an overappreciation in our society of the role played by innate differences among individuals in determining what can be learned and how much can be learned, and that overappreciation is coupled with an underappreciation of the power of training, practice, and experience," they write. This combination "can lead individuals to assume that there are certain limits on what they can learn."

As I've noted, there is a multitude of reasons why the human potential all around us goes unrecognized and unrealized. Many, perhaps most, have to do with systemic injustices—societal failures that it's incumbent upon us to address directly. But these inequities are often only compounded by the cognitive roadblocks we've set in front of learners, and the cognitive blind spots that determine how students are sorted.

I hope it's now clear that we don't have to live with these stumbling blocks and blinders. There are steps we can take in the here and now to clear them away. The same steps might not be the right choice in every situation, and different approaches may appeal to different practitioners. But the overarching point remains undeniable: We can learn, and teach, differently. No longer beholden to a nineteenth-century idea of the learning mind, we can keep pace with science's multiple, ever-advancing cutting edges.

In fact, not only can we put cognitive science discoveries into practice, but I believe we're ethically obligated to do so—posthaste. As in the rare clinical pharmaceutical trial that goes so well that the researchers pause the study and start handing out the lifesaving drugs to the control group, the benefits of cognitively friendly instruction are so profound that inaction at this point would be tantamount to malpractice. We have knowledge that can save students from the educational winnower and help them realize a lifelong love of learning. It's time to put that knowledge to work.

Part Two

MIND AND HAND

- VI -

VOYAGES

In our climb up through the layered disciplines of cognitive science, we've only dealt in passing with how to apply their findings to learning and teaching. Now it's time to make the same turn that all students of engineering must learn to make: to gather up our abstract scientific knowledge and put it to work in the real world.

This transition is anything but straightforward. For one thing, the real world places different demands on engineers than on scientists. In the case of cognitively user-friendly learning, such seemingly mundane details as whether a given pedagogical tactic can be realized at scale, let alone cost-effectively, take on prime importance. Even the most amazing instructional idea can never live up to its transformative potential unless we find a way to open the floodgates and let large numbers of people experience it—and not just in their youth, but throughout their lives.

Meanwhile, the logical traps that bedevil basic science become, if anything, more perilous in its application. When scientists oversimplify, it's usually by creating a model whose scope is too small: failing to account for all the causes responsible for an observed effect. Engineers, meanwhile—scientists inverted, in a sense—tend to oversimplify by making scientific models too large: stretching them to predict real-world effects beyond what their logical skeleton can bear. Reductive thinking in science is never good, but only in engineering does it become truly dangerous, manifesting in bridges that collapse, financial instruments that destabilize economies, drugs with unacceptable side effects, winnowers that waste grain.

In any quest to apply findings from our cognitive high-rise, then, the first step must be to understand what cognitive science research *doesn't* tell us. Perhaps most glaringly, when cognitive scientists attempt to clarify how "the brain" or "the mind" works, they're generalizing: studying many individuals (even multiple species) in an effort to paint a picture of a generic human brain, a generic human mind. In reality, however, there is no single, typical human brain. And so for any truly inclusive, cognitively friendly vision of education to work in the real world, it will have to be flexible. Wherever possible, it must tolerate individual variation in student interest, motivation, prior knowledge, speed of learning in a given subject, and far more.

Scalable, flexible learning that is optimized for cognition: any one of these three engineering demands is challenging enough to realize in a school. Taken all together, they raise a stark question: Are the reforms we're mulling even achievable within our inherited educational institutions? Or will we need new institutions, built from the ground up, to make these virtues possible?

—

In fact, there's quite a bit that an enterprising reformer can pull off, even while confined by traditional educational structures. Part of what makes the Florida International University College of Law story so remarkable, for instance, is that it achieved its profound reversal *within* the familiar trappings of law school education: pontificating professors, semicircular lecture halls, final exams, and so on.

MIT, too, has tested how much instructional change the larger academic superstructure will tolerate. Fittingly, a literal space explorer is responsible for the Institute's initial giant steps down this path. John Belcher was part of the team that built the plasma-detection instruments on the *Voyager 1* and *Voyager 2* spacecrafts and, during their flybys of Jupiter, Saturn, Uranus, and Neptune in the 1980s, served as those instruments' principal investigator. Today, he spends much of his time poring over the data now returning from the probes as they venture into interstellar space. In between then and now, however, he turned his gaze to the challenges of physics education—which, as lead lecturer for MIT's largest course, Physics II, he found he couldn't ignore.

At a conference of physics educators in 1996, Belcher encountered several different tactics for improving on the standard lecture-and-recitation model. Like so many other frustrated physics professors, he had observed that even top-tier students, even after extensive physics coursework, still tended to come up with intuitive but wrong answers to physics questions. The problem, he was told by scholars who had studied their Piaget, had to do with the sorts of explanations that students construct early in childhood for how the physical world works. Upon encountering natural phenomena like momentum, friction, and weight, children readily form personal theories that are correct enough to serve their purposes in daily life, but fail to hold up in the physics classroom. Correcting such misconceptions required a deep remodeling of very old schemata, densely interconnected with memories acquired over the course of more than a decade. Such deeply held rules couldn't merely be reasoned away in the space of a lecture. No, they needed to be eradicated roots to branches—an approach, he was surprised to hear, few instructors had tried.

Already, however, that was beginning to change. Up the river at Harvard, the physicist Eric Mazur had begun experimenting with the first of two major innovations he would contribute to classroom practice around the world. In his system of "Peer Instruction," as he named it, he encouraged students to explain confusing physics concepts to each other. "You're a student and you've only recently learned this, so you still know where you got hung up," he later recalled. "Whereas Professor Mazur got hung up on this point when he was seventeen, and he no longer remembers how difficult it was back then."

Peer learning acted sort of like a time travel device: allowing professors to climb down one student's schematic tree to meet other students at their level. There was a problem, however: Although the physical shape of traditional lecture halls exposed all students to equal quantities of professorial wisdom, it prevented them from easily speaking to one another.

Soon enough, however, Belcher discovered that a handful of pioneering physics departments—most notably at the Rensselaer Polytechnic Institute and North Carolina State University—had been testing out a solution. "Studio Physics," as it was known at

RPI, combined aspects of lectures, problem-solving recitations, and laboratory work into a single block of time, and replaced the one-way communicative architecture of the lecture hall with small round tables and an open floor plan—perfect for instituting a version of Mazur's Peer Instruction.

Even better, the setup moved demonstrations down from the lecture stage to the students' tables, where they could physically experience the objects and processes involved. To physics lecturers of the old school, such touchy-feely accommodations might have seemed unnecessary. But for a teacher hoping to undo physics misconceptions formed in childhood, the approach made perfect sense. According to the theoretical perspective underlying the approach, known as *embodied cognition*, the brain is not functionally limited to the neurons living in the skull. Rather, for all practical purposes, it stretches out in neural networks running throughout the body, which play crucial roles in early development and, in adulthood, continue to factor into the encoding, retrieval, and modification of memory. According to this framework, experiencing a physics concept through one's own hands would create an overlapping yet distinct set of stored associations as compared to reading about it, or observing it from a distance. By encouraging such tactile experiences, studio physics could reinforce students' understanding of concepts—and, better yet, enable professors to travel even further back through students' learning histories to address misconceptions where they were created, in childhood.

Before Belcher and a group of like-minded educators could build MIT's own version of studio physics, however, there were still a handful of architectural problems to overcome. For instance, in studio physics–type classrooms, Belcher often observed a large gaggle of students crowding around a single whiteboard, solving problems. A bigger, better, tech-enabled classroom was needed—which would be possible only with ample money and full buy-in from the Institute. Belcher demanded and obtained both. As Peter Dourmashkin, Belcher's collaborator and eventual successor in this effort, recalled: "He got the grants based on his reputation, based on his experience at MIT. He was able to initiate the project, and he built a team."

At the time, there were a few major sources of grant money floating around, in search of big promising ideas for education reform.

"A lot of other people got some money," Dourmashkin said. "But only TEAL"—as Belcher's Technology Enabled Active Learning system became known—"made it into the classroom."

TEAL AND BEYOND

In space, there is no up or down. In a TEAL classroom, there is no back or front.

As a result, if you were to sneak into a TEAL class at MIT, you'd likely behold an apparently random system that, upon closer inspection, betrays hidden organization, like the movement of heavenly bodies. The professor—these days frequently Dourmashkin—stands in the middle of the three-thousand-square-foot space, surrounded by thirteen tables, at which sit nine students apiece, divided into groups of three. Among those tables pirouette a technical instructor, a grad student, and six undergraduate teaching assistants, while on their Formica surfaces unfolds the music of the spheres: classical, Newtonian physics demonstrations. (Or, in Physics II, electromagnetic demonstrations involving, for instance, tabletop Faraday cages.) Lining the walls, you'll find whiteboards and projector screens. Ceiling-mounted video projectors allow the professor to multiply any group's whiteboard thirteen times over, so that the entire class can see it. At other times, he or she may project advanced visualizations of complex electromagnetic fields, generated using student data. (Belcher, who had developed an abiding interest in electromagnetic field shapes dating back to his early *Voyager* experiences, built the underlying software himself.) Finishing the picture, anachronistically, are the desktop computers on each table: holdovers from the classroom's founding, a few years before most students brought laptops to class.

Back in those days, when MIT made the sudden switch from the lecture-and-recitation model to TEAL, the upheaval—both architectural and cultural—was fresh, and skepticism emanated from all quarters. "What we learned right from the start was there were really three communities that you have to deal with extensively," said Dourmashkin: "administrative, faculty, and student cultures. And if you neglect any of those, you run into huge problems." Students, for instance, chafed at being forced to attend TEAL's mandatory class sessions. (MIT students, perhaps more than most, have a reputation for skipping lectures that cover the same ground as a course's textbook.) They also feared the group work involved. To Belcher as well as Dourmashkin, a staunch Peer Instruction believer who had collaborated with Eric Mazur, the group-interaction aspect of TEAL was non-negotiable. It was necessary not only to realize the benefits of peer-to-peer teaching, but also for students' development as future members of laboratories and collaborative workplaces. The promise of collective work, however, seemed to set off a Pavlovian trigger in students. "Their idea of group learning was from high school," Dourmashkin said, where, as high achievers, many had grown accustomed to getting stuck with the bulk of the work.

If students proved restless when TEAL began in earnest, some faculty members found their cortisol levels spiking to equivalent heights. "There were some faculty members who were upset that they weren't going to be lecturing anymore, and they did a lot of kind of getting the students to petition against TEAL," said Dourmashkin.

"We feel that the quality of our education has been compromised for the sake of 'trying something different,'" read the 2003 petition. "It should not be forced upon the majority of the student body." In its coverage of the issue, the student newspaper included critical quotes from prominent professors. TEAL had "a whole spectrum of problems," said one physicist. "Many students are really angry."

And so, for a moment, TEAL seemed in real danger. Had TEAL changed too much about traditional physics education, too quickly, and with too little respect for the institutional norms and bylaws surrounding it?

Ultimately, with an outspoken set of students and faculty in revolt, it was a group of administrators who saved the day. The larger physics department understood that the program would experience atmospheric turbulence on its way to stable orbit, and came through with the one resource that mattered above all others: patience.

A few years were exactly what the TEAL team needed. Training teachers, which had taken a back seat during the program's initial launch, took on primary importance. Some were so accustomed to delivering traditional lectures, Dourmashkin said, that "they wanted to take out space in the classroom and make a tilted floor."

Retraining them "was number one," he said. "Two, we had taken data." Within a few years, they could "show that the students in TEAL had higher learning gains than students who were in the lecture."

From the start, TEAL had been set up not just as a classroom intervention, but also, in the tradition of Dewey's Laboratory School, as an experimental one. By 2006, the results from the experimental effort, led by Yehudit Judy Dori, an education researcher at the Technion, in Haifa, Israel, were rolling in. The experiment relied on a variety of assessment techniques, but perhaps most important was a testing regime, posed to students before the course and after, featuring the sorts of conceptual questions that tend to expose inert knowledge.

By the end of the semester, TEAL students answered those tricky conceptual conceptions twice as well as those who went through the old lecture format. The TEAL failure rate, meanwhile, was less than half that of the lecture. Eighteen months later, TEAL students still outperformed the traditional class.

TEAL's greatest beneficiaries were women. To the shame of introductory physics courses around the world, men have historically outperformed women—an indication of structural bias, not aptitude, that contributes to continued gender imbalance in professional physics research and engineering. "I think physics was an older kind of gatekeeper style of learning," Dourmashkin said. "This was driving a lot of women out of physics, and I think the peer collaborative environment changed that." The findings tracked closely with similar research from a Swarthmore-based team that included Mazur, which had shown that Peer Instruction in various contexts boosted both men's and women's academic performance, but women benefited more. In the specific case of TEAL, in some semesters women accounted for all six of the laboratory's undergraduate teaching assistants. That set a tone: Whenever a new group of students walked into such a classroom for the first time, "the first thing they saw," said Dourmashkin, were the female students who would help lead the show that semester. "I think that delivered a really interesting message to everybody." Today, in TEAL-based physics classes at MIT, the gender performance gap is gone. "If you look at final exam scores," he said, "the average is almost identical."

SPACETIME

Since its invention in 2001, TEAL has spread around the world, and up and down the age spectrum. The results keep rolling in. In Taiwan, for instance, the introduction of TEAL didn't just boost high school students' exam scores, but it also added to their self-reported interest in physics in general, and participation in extracurricular science programs. At the University of Kentucky, the introduction of a fifty-four-seat TEAL classroom reversed the gender gap and then some: women went from underperforming men by 5 percent to outperforming them by 10 percent. Without the right instructor, the classroom itself isn't a wholesale guarantee of success, as a study of college-age students in Taiwan showed—which echoed MIT's early experience with untrained TEAL instructors. The same study did, however, reaffirm the findings from other countries: In the new TEAL classrooms, women thrived.

Looking back, TEAL's early success in keeping women from being informally winnowed out of careers in physics and engineering is one of the most inspiring results the team could possibly have asked for. The fact that TEAL has since spread, thanks in great part to MIT's bully pulpit, and is ushering more women into the fold, only confirms its status in my mind as an unmitigated success story.

But perhaps the most intriguing part of that story is that TEAL accomplished so much in those early years with one hand, technologically speaking, tied behind its back. Initially, the "Technology" component of Technology Enabled Active Learning referred to matters having to do mainly with *space:* video cameras projecting a single group's whiteboard around the room; complex field simulations tied to desktop experiments.

Far more revolutionary, however, was TEAL's follow-up technological wave, which pushed the boundaries of *time.*

For one example, take Eric Mazur's second major contribution to classrooms, for which he has become justly famous: the Personal

Response System, or "clicker." The technology will be familiar to anyone who has ever seen the game show *Who Wants to Be a Millionaire:* a handheld radio device that allows a large studio audience—or classroom—to answer multiple-choice questions en masse.

Clickers, which have long since become a common sight in large university lecture halls, allow professors to gauge student comprehension of a given topic before moving on to the next. In retrospect, the idea seems almost obvious. After all, teachers have always relied on informal polling techniques to determine when to move on: We pose questions; we look to see how many heads are nodding. All clickers do, in theory, is beef up such methods' statistical accuracy.

But the technological premise of clickers is more profound than that. It hints at a possible change to the status quo that is not incremental, but qualitative. In its small way, in fact, the clicker represents an act of rebellion against a rule of school predating even Thorndike: that students should keep up with their teacher. With clickers in hand, students help set the pace.

It is joined in its revolt by another technique used in many forward-looking classrooms, TEAL included: "blended" or "flipped" learning. In such approaches (the former tends to be a measured version of the latter) students take in some or all lecture content in video form prior to class, and then complete some or all of their homework during classroom hours, with instructors close at hand and ready to help. As far as classroom trends go, nothing is quite so trendy as "flipping." In one 2016 survey, 55 percent of college and university instructors said they were in the process of flipping or blending at least one of their courses. In the case of MIT's TEAL classes, Dourmashkin and the other instructors make the requisite video lectures themselves, not in an auditorium but a black-box studio, separated from the camera by a well-lit pane of glass. Onto this surface, they write and draw diagrams in Day-Glo marker as they lecture, which, because the video can be mirror-reversed after the fact, allows them to face their students while writing—a professorial first.

Taken together, both clickers and flipping classrooms represent some of the biggest changes to classroom practice since chalk met board, and the reason why they feel so profound is devastatingly simple. They challenge the most fundamental assumption of contemporary mass education: that learning must be parceled out not in units of knowledge, but units of time.

From early youth up through high school, college, and even into graduate school, our progress as learners is measured in the number of hours we spend in chairs in classrooms, the number of weeks those classes take to run their course, and the number of years it takes to earn a degree. Courses trundle along at the same pace for students pulling A's and C-minuses alike—frequently to the boredom of the former and terror of the latter.

Clickers and flipped learning practices occupy a unique position in modern education because they break the spell of time-centric education without violating school's many other inherited institutional structures. With a clicker on your desk, you can vote to slow down or fast-forward your professor's delivery; watching a video lecture at home, you can do the fast-forwarding and rewinding yourself. However, your course still fits into a larger, time-centric arc: at the end of the semester, you will still receive a grade representing knowledge accrued *as a function of time.* That grade, meanwhile, will still fit into your transcript, just as the course fit into your semester's schedule, just as your classroom—even a new, high-tech TEAL classroom—fits into the physical architecture of your school.

All told, a given course may play fast and loose with space and time, but, like an episode of *Monty Python* airing between news programs, it's just a modular bit of wildness shoehorned into an otherwise staid structure.

Engineers have an expression for the expected performance limits of a given air- or spacecraft: the *flight envelope.* To exceed those limits is to *push the envelope*—an expression that has long since escaped into the vernacular. The MIT TEAL program pushed the envelope of the traditional classroom model so far beyond its design specifications that it nearly crashed after liftoff. (Instead, all it broke was records.) But now, it's not clear how much more modification it can handle.

The TEAL diaspora is now part of MIT's worldwide legacy. But at the same time, one can't help but wonder what we could accomplish without the program's inherent limitations: time-centrism, the encroaching administrative superstructure, and—not to be overlooked—TEAL's cost, which can be prohibitive for even well-heeled schools.

One need not wonder too long, however. In fact, in the early 1800s, the world very nearly embarked on a very different system of mass education—an exceedingly economical one, as it turns out—in which knowledge, not time, reigned, and which had its own unique rules of school to reflect that fact.

As I'll discuss in the chapters ahead, learners are only now starting to find their way back to similar approaches via a variety of independent paths, aided in many cases by technologies of the sort used in TEAL classrooms. The vast majority have no idea that societies around the world have seen this idea before—that they tried it and, despite its theoretically impressive pedagogical power, discarded it. Now, as we find our way to similar strategies, it's essential to understand what happened, and why. We must travel back to see the world as the young Scotsman Andrew Bell saw it—before he changed the very history of education, and before educational history forgot him in return.

THE MALE ORPHAN ASYLUM

In June 1787, when Bell stepped off his ship into the Indian port city then known as Madras, he merely intended to stretch his legs for a day or two before re-embarking for Calcutta. He ended up walking into a crisis—and staying a decade.

In the 1780s, India was suffering the spread of colonial rule, mainly under the British, and that stranglehold was only growing in strength. Earlier in the century, France had been the dominant colonial power on the subcontinent, but in India, like just about everywhere else, the French came to blows with the British, and by 1783, the latter had emerged as the clear favorite.

Looking back on this period, one aspect of the occupation that doesn't get nearly enough attention was the large number of Indian women who became pregnant with the children of colonial soldiers, whether due to rape or coercion or merely the prospect of a marginally better life under oppressive rule. Many of those soldiers then died or were called away to serve elsewhere, leaving these women and their children behind. Despite the fact that most of these children had living mothers, the colonial government considered them to be military orphans, and determined—often, it must be said, against the will of the mothers—that something had to be done about their housing, feeding, and education.

This challenge would become Bell's primary concern during his time in Madras, the city later (and more correctly) known as Chennai. As he stepped off the ship at age twenty-eight, he was crackling with unrealized potential. He'd traveled abroad once already to seek his fortune: to the American colonies, which had declared independence while he was there. He barely lived to tell about it. Right before the Revolutionary War found its way to Virginia, where he'd been living and working as a tutor, he fled on a Britain-bound ship. On his way to the docks, he even passed the Marquis de Lafayette—the French general who came to the aid of the Americans—headed in the opposite direction. Bell's ship struck ground off the desolate shore of Nova Scotia, however, and for several weeks, the ship's crew and passengers camped on a snowbound island, freezing in the Canadian spring. Finally, after being rescued by a whaling ship and recovering in Halifax, Bell returned to his native St. Andrews, where he nearly died twice: once due to a sore throat so severe he couldn't swallow for three days, and once in a pistol duel, which happily resulted in no injuries.

Fate, having evidently realized that killing Bell wasn't worth the effort, decided instead to take his side. Bell's father had found himself the swing vote in the election that would determine the town's representative in Parliament, and the resulting winner promised to take an interest in the young Bell. This initially took the form of advice not to go into the Church of England, which Bell, who was ordained a priest in 1785, ignored. Two years later, more fruitfully, Bell's patron came through with an honorary MD from St. Andrews University, a berth on a ship bound for Calcutta, and letters of introduction.

These letters were of no use in Madras, but by the time the ship arrived there, Bell had ingratiated himself with the captain and other dignitaries onboard by giving lectures and aiding in naturalistic observations. Armed with a whole new slew of introductions, the young Reverend Doctor made something of a splash in Madras society. He began delivering paid lectures that soon amounted to major social events. Bell demonstrated for his adoring onlookers the principles of electricity, created India's first-ever artificial ice, and built the subcontinent's first hot-air balloon. As his earliest biographer recounted: "It was of no great dimensions; for as the assistant did his part badly, and the thing failed, Dr. Bell (in his own words) threw it in a passion from the verandah. After which the heat of the sun rarefied the enclosed air, and the balloon mounted in grand style."

Bell's growing legend in Madras caught the attention of the backers of the Male Orphan Asylum, who were in the market for a suitable superintendent. He readily accepted the job. The asylum would take several more years to get off the ground, however, during which time Bell continued to lecture for cash. Even more financially rewarding were his chaplaincies. Upon arrival in Madras, he quickly became a chaplain of a local regiment, then another, until he held a total of eight chaplaincies at the same time, which added up to generous pay for relatively little work.

In 1789, the East India Company opened the Male Orphan Asylum in an enormous structure known as the Egmore Redoubt, which had been used to house gunpowder and had, in its relatively young history, already been blown up twice. (One of Chennai's train terminals now stands at the site.)

From the start, the asylum was a necessarily lean organization. Its main funding sources were an allowance provided by the East India Company, charitable donations, and fines imposed upon British soldiers for drunkenness. These (especially, one assumes, the last) added up, but weren't equal to the challenge of housing and educating the local "orphan" population of 230 boys. When Bell first joined, the school was teaching, feeding, and sheltering just 20 of them, although this number soon increased to 100 as funding sources began to congeal. Against the advice of his friends, Bell, who was not yet wealthy, volunteered his services gratis, surviving on the combined income of his chaplaincies and lectures.

Initially, Bell's attention was so preoccupied with such concerns as feeding, clothing, and inoculating his students against smallpox that he had little time to think about their education. The school's original uniform and bedding proved insufficiently warm in the rainy season, and its board insufficiently nutritious. Cold and hungry, most students became infected with roundworms, and many with measles. At one point, nearly a third of the student body could be found in the local hospital, and a medical report characterized many of the children as "so puny, that it must be great care indeed which could save them." During the school's first three years, Bell lost four students to disease, including smallpox. It could have been worse: The measles "proved less fatal than had been apprehended," one biographer reported.

Educating the children added a whole separate order of difficulty. When he took control of the asylum, Bell assumed authority over two assistant teachers, known as "ushers," and one headmaster. From the start, Bell was astounded by their inability to teach, especially when it came to the youngest students and the alphabet. But every time he managed to hire a suitable new assistant teacher, a better position for that person seemed to open up elsewhere. And in the meantime, it seemed that whenever he suggested changing this approach or that, his remaining assistants took umbrage and dragged their feet.

It was with all these frustrations whirling that Bell took a morning horseback ride along the Madras coast. He happened to pass an open-air Indian school, where he observed something curious—to his eyes, anyway—taking place. Tracing figures in sand on the beach, older children appeared to be teaching younger children to write. He stared, and then took off—at a cinematic, sand-spraying gallop, I like to imagine—crying "Eureka! I have discovered it!" as he went.

He wasn't the first, nor would he be the last, colonizer to claim to have "discovered" something that a society had been doing for a long time. What he'd stumbled across was likely a common vernacular school, where education was conducted in one of the local languages—likely Tamil, Telagu, or Marathi—as distinguished from more rarefied schools, often conducted in Sanskrit. In South India in the eighteenth and nineteenth centuries, a common instructional technique at such vernacular schools involved one student writing figures from, say, a table of weights and measures, reading them aloud, and a group of his fellows following his lead and rhythmically reciting it. It was probably this tactic or something very similar, and the fact that the boys were enacting it by tracing figures in the sand, that made Bell do his best Archimedes impression and cry "Eureka" all the way home.

THE SCHOLAR FINDS HIS LEVEL

Back at the Male Orphan Asylum, Bell instituted changes. He started by providing each of his youngest students with a small tray of wet sand. He told his adult teaching assistants to instruct them to write their letters in the sand using their fingers, which created a reusable writing surface that would never run out of space, unlike the expensive copybooks then in use. "It engages and amuses the mind," Bell later wrote, "and so commands the attention, that it greatly facilitates the toil both of the master and scholar."

Despite these benefits, Bell's assistant in charge of the youngest children told him that such a departure from tried and true methods would be impossible. Frustrated, Bell decided to borrow the other surprising practice he'd observed at the beachside school: mutual instruction. He began paying one of his older students, one John Frisken, a small fee to teach the youngest children their letters using their sand trays. In this very early vindication of Peer Instruction (at least according to Bell's sometimes hagiographic early biographers), Frisken soon easily outdid the adult teaching assistant, and Bell expanded the practice of paying older students small sums to teach younger ones. What worked in one class proved successful in others, and Bell soon had regiments of boys working under each other's tutelage.

"The school is arranged in six or eight classes," Bell later wrote, in a pamphlet he would use to spread the word about his system to the Anglophone world. Bell's students were not grouped into classes in the sense that we now understand them, where every student is fed information at the same rate. "No Class is ever retarded in its progress by idle or dull boys," he wrote. Rather, classes remained static while learners moved *through* them at their own pace, like martial artists earning new belts, or skiers advancing from blue squares to black diamonds. In Bell's system, a boy at the top of his class was given the opportunity to jump up to a higher class, where he would start at the bottom. He would be given a few days to climb up to the middle, and if he failed to do so, he would be demoted to his earlier class, where he would remain until ready to try to climb again.

Meanwhile, any boy who repeatedly failed in his daily lessons would slide back one class, where he would sit at the head of the group. If his performance continued to slip, "he is doomed to permanent degradation," Bell wrote. "But, if he maintain a high rank, he is allowed to resume his original Class on a new trial; when it often happens that, by redoubled exertion, he can now keep pace with them."

Within each class, Bell arranged his students into pairs of pupils and tutors, the latter of which received additional instruction outside of school hours. "Mark, at the outset, how many advantages grow out of this simple arrangement," he wrote. "First, the very moment you have nominated a boy a Tutor, you have exalted him in his own eyes, and given him a character to support, the effect of which is well known. Next, the Tutors enable their Pupils to keep up with their Classes, which otherwise some of them would fall behind."

This last point, to Bell, was of the utmost importance. To this day, falling behind in class remains a self-compounding problem, with one unlearned fact leading to another, then another. "This," he wrote, "is the reason why some boys in most schools are declared incapable of learning." The blame didn't rest with them, he argued, but with their instructors. "It is you," he wrote, addressing their hypothetical teachers, "who do not know how to teach, how to arrest and fix the attention of your pupil: it is not that he cannot learn, but that he does not give the degree of attention requisite for his share of capacity."

There were other advantages to Bell's system. Teaching a concept seemed to help the tutors learn the material more effectively than if they had taken it in only passively, an experience Bell remembered from his younger days as a tutor and which, years later, Robert and Elizabeth Bjork would recognize as an example of retrieval practice. The system also relied on quick, fifteen-minute lessons and hands-on work (including a requirement that every child who graduated from wet sand trays to paper first had to construct his own pen).

To keep the boys on task, Bell instituted a complicated system of rewards, which he believed motivated students more effectively than the fear of punishment. He reserved classroom discipline solely for unruly behavior, not academic underperformance, and limited punishment to such measures as detention and writing assignments, which were doled out by a jury of the accused's peers. Corporal punishment, youthful memories of which haunted Bell, was verboten.

As the full scope of Bell's changes came into view, his headmaster and assistants rebelled. By the time the headmaster handed in his resignation, Frisken was eleven years old and running a third of the school. Bell did manage to retain some of his assistant teachers, however, who became engaged less with teaching than with the task of ensuring that the boys didn't abuse their newfound responsibility over one another. "Such interference prevented all that tyranny and ill usage from which so much of the evil connected with boarding-schools arises," wrote one of Bell's biographers.

That probably wasn't entirely true; bullies tend to find a way. But still, by Bell's account, anyway, the system appears to have worked on levels both academic and—crucially, for our purposes—financial. Expenses, on a per-student basis, fell substantially following the institution of Bell's new system. Initially, the school's overall per-student expenses amounted to roughly ten rupees per month—and that was before Bell demanded improvements in the students' diet and dress. His new system, however, together with other efficiencies he'd discovered along the way, such as owning milk cows instead of paying for milk, meant that the asylum could now house and educate each student for a little more than six rupees—a cost reduction achieved apparently without any compromise in terms of dress, room, or board. "On no occasion, and on no account, had ever any deduction been made from the allowances of the boys," Bell claimed.

In the asylum's full-blown form, its student body of two hundred was taught almost entirely one student to another, with none of the students older than fourteen. By 1792, Bell was crowing about his new system in letters to friends in Britain: "Every boy is either a master or a scholar, and generally both. He teaches one boy, while another teaches him. The success has been rapid." Mothers, Bell wrote, had initially mourned the loss of their sons to the boarding school—and for good reason, since many were coerced into sending them. But under his new system, he claimed, they "ply us now with every species of importunity to have their younger children admitted." Living British officers, meanwhile, fought to have their sons educated alongside the asylum's orphans. "We have already more than thirty boys, white and blue"—the term for a child of mixed parentage—"of this description, though they are subjected to the very same treatment, dress, discipline, and diet, as the poor orphans," wrote Bell. "This I consider as the best commendation of the Asylum."

—

Bell set sail for Britain in 1796 because he'd been told the climate there would benefit his flagging health. Returning with £25,000 and an East India Company pension to his name, he proceeded to go through the motions of settling down. He bought a sizable, rent-bearing Scottish estate; accepted a rectorship in Dorset (where he successfully vaccinated his congregation against smallpox); and, "at the not immature age of 47," noted one biographer, he married one Agnes Barclay, the daughter of a local minister. With his roots once again established in British soil, he spread his branches. In 1797, he had printed one thousand copies of a book describing his achievements in Madras, and handed them out to every influential person he knew.

"You may mark me for an enthusiast," Bell wrote in a letter to the book's printer, "but if you and I live a thousand years, we shall see this system of education spread around the world."

By the next year, England's first Madras-style school appeared. A few sprung up, or were converted to the Madras system, annually for several years after that. Enthusiasm for the system soon began to multiply thanks in no small part to endorsements from such quarters as the archbishop of Canterbury, a number of influential lords, and even the poets William Wordsworth, Samuel Taylor Coleridge, and Robert Southey—the last of whom wrote Bell's first major biographical treatment, published in 1811. That same year, the National Society for the Education of the Poor in the Principles of the Christian Church, a powerful group with a self-explanatory title, was founded and set up all of its schools according to Bell's system. By the time of Bell's death in 1832, this organization was running the Madras system in more than twelve thousand schools in Great Britain and its colonies. And that was just the one organization: Madras also found its way into other systems of charity schools and was exported far abroad, from the Caribbean to Oceania and

South Africa, even to Russia. It could even be found in snowy Nova Scotia, where Bell had been shipwrecked decades earlier.

Although Bell's system overtook most of the English-speaking world and then some, there were exceptions—including, critically, the young United States of America. The problem wasn't that Americans were opposed to the system. Rather, it was that the United States was the stronghold of Bell's greatest rival, who had started out as his biggest supporter.

—

Bell and his wife first met Joseph Lancaster in 1804 at their home in Dorset, where they greeted the young man warmly. Lancaster, twenty-five years Bell's junior, was developing an educational scheme that would ultimately share a number of features with Bell's, including a large, one-room schoolhouse, instruction that passed from one pupil to the next, and the use of trays filled with wet sand. Bell and Lancaster tolerated each other in those early years, but a schism soon formed: Bell was loyal to the Church of England while Lancaster, a Quaker, insisted that his schools' Christian instruction remain nondenominational. As a result, Bell's version of "monitorial" education, as the systems became known, retained the advantage wherever England held sway, while Lancaster found success elsewhere, including Gran Colombia (now Venezuela), Mexico, and the United States. Lancaster's custom was especially welcome in the Quaker city of Philadelphia and nearby New York City. There, he found a champion in DeWitt Clinton, the most powerful politician in the state.

In the young United States, there was a particular urgency animating the drive for mass education. The American republic was still new, and its system of government untested. Today, a representative government may seem like a fairly stable way to run a country, but at the time, there hovered the ambient worry that the forces of anarchy and monarchy would prevail. The most obvious way to keep them at bay was through education. As Thomas Jefferson argued in 1778, at the height of the Revolution, "those entrusted with power have, in time, and by slow operations, perverted it into tyranny," and that "it is believed that the most effectual means of preventing this would be, to illuminate, as far as practicable, the minds of the people at large."

It was in this spirit that Philadelphia and New York began to look into establishing modest systems of public schools for their poorest kids. This push predated modern tax-funded education, which meant that the charitable organizations behind the schools were constantly scrambling for money. As a result, when Lancaster appeared, promising maximal education per dollar spent, they rolled out the red carpet. In a speech given at the opening of a Lancasterian school in New York City in 1809, DeWitt Clinton waxed rhapsodic: "When I perceive one great assembly of a thousand children, under the eye of a single teacher, marching, with unexampled rapidity and with perfect discipline, to the goal of knowledge, I confess that I recognize in Lancaster the benefactor of the human race."

Although the early 1820s were a heady moment for Lancasterism, even then gears could be spotted flying out of the supposedly flawless machine. True, whenever DeWitt Clinton visited one of Lancaster's schools, he beheld scenes of perfect concord, but then again, he was one of the most eminent Americans alive. As one Albany schoolmaster put it in an 1818 letter to him, "There is as much waywardness in the youth of Lancasterian schools as of any other. Oft does the proud spirit of Fredonia's sons rise in mutiny against the authority of one whom they consider at least no better than themselves;—but I presume you never saw that system in operation, but when DeWitt Clinton was present. When such a name was but whispered round the School, what stillness! what subordination! what assiduity!"

Indeed, discipline was enough of a concern in Lancaster's schools that their founder experimented with a baroque system of almost gleefully creative punishments to keep his students in line. Though technically considered non-corporal at the time, these would send parents today in search of a police officer, lawyer, and therapist, in that order. A child who broke the rules one too many times could wind up with a "wooden log round his neck," Lancaster explained in an 1803 treatise. "While it rests on his shoulders, the equilibrium is preserved; but, on the least motion one way or the other, it is lost, and the logs operate as a dead weight upon the neck. Thus, he is confined to sit in his proper position." That wasn't all. A student who broke the rules could also have his legs attached together with wooden shackles: "Thus accoutered, he is ordered to walk round the school-room, till tired out." If that didn't work, Lancaster suggested tying the student's left hand behind his back, or his elbows together. When offenders erred in pairs or groups, they could be "yoked together sometimes, by a piece of wood that fastens round all their necks: and, thus confined, they parade the school, walking backwards." For repeat scofflaws, Lancaster reserved his most bizarre and degrading punishment. "Occasionally boys are put in a sack, or in a basket, suspended to the roof of the school, in the sight of all the pupils, who frequently smile at *the birds in the cage*"—emphasis his.

—

Given this unattractive glimpse into Lancaster's personality, it may come as little surprise that he alienated allies as quickly as he made them. A major reason he left Britain for the United States, in fact, was that rumors had surfaced that he'd whipped some of his young monitors for his own amusement, and his English benefactors effectively gave him the boot. This group, the British and Foreign School Society, acquitted themselves perfectly capably in his absence, working with missionaries to establish Lancaster's schools abroad. They soon appeared throughout the Caribbean, in Egypt, Malta, Australia, Sierra Leone, Madagascar, the Cape of Good Hope, and even gained a foothold in India and Ceylon (now Sri Lanka). Attempts to introduce the system in Germany, Switzerland, and Holland faltered, but it achieved meteoric growth in Latin America. Demand for public education accompanied the Latin American revolutions of the 1820s, and students braved the possibility of being impressed into passing armies as they walked to their monitorial schools, which popped up in many major South American cities as well as in Panama and Mexico.

In the United States, meanwhile, Lancaster made short work of estranging his friends as he bounced from state to state, always in search of a new position befitting his genius. A timely missive from South America arrived just as he was wearing out his welcome in Baltimore, and he set off for Caracas bearing the invitation of Simón Bolívar, the revolutionary hero. Upon arrival, Lancaster was given a welcome befitting a visiting dignitary, and, when he met and married a woman there—the widow of a long-lost acquaintance—Bolívar delivered the wedding toast. El Libertador and Lancaster soon had a falling-out, however, and the latter found himself back on the road with his new family in tow: to Trenton, to Montreal, to Philadelphia, always leaving a trail of unfinished projects and angry benefactors in his wake. "Lancaster was always planning great things and never doing them," writes the education historian Carl Kaestle. "He probably petitioned more famous people for hand-outs than any man of his day; the list includes Roberts Vaux, Gulian Verplank, Andrew Jackson, Martin Van Buren, DeWitt Clinton, and George IV."

In 1838, Lancaster was trampled to death by a runaway horse in New York City. Bell had died six years earlier at the age of seventy-eight, and was interred with great ceremony in Westminster Abbey. Even before their founders' deaths, however, both Bell's and Lancaster's systems had begun a long slide into disuse. By the middle of the century, cities around the world were replacing monitorial schools with institutions that would feel far more familiar today, with classes organized by age, not proficiency, and self-contained classrooms taught by adults, in which every student was expected to proceed at the same pace.

There are a few plausible explanations for the worldwide downfall of monitorial education. In the United States in particular, Lancaster's "pauper" schools took on something of a stigma. As one unnamed New England critic put it as early as 1832: "The sole merit of [Lancaster's] plan is, that it saves money . . . I can easily imagine that such a school may make excellent sailors and soldiers; for they are expected to be automatons. But for republicans, for freemen, for self-controlling, and elevated masters of their own destiny—it is not the place." The glee with which Lancaster approached punishment only reinforced the idea that his schools might be acceptable for other people's kids, but were hardly a place you'd aspire to send your own. "We certainly can afford something better," the critic wrote.

By the 1840s, it was becoming clear that that "something better" would consist of self-contained classrooms taught by professional teachers, an approach championed by Horace Mann, Massachusetts' first education secretary and an influential advocate for public schools. In a preview of the economic and cultural changes to come in the Progressive Era at the turn of the century, established social structures had been falling away as interstate commerce, abetted by new railroads and canals, undercut many a local, proprietor-owned business. In an effort to take up some of the resulting slack in both social authority and social services, states embarked upon a frenzy of institution building, establishing poorhouses and hospitals, mental asylums and prisons. Early in his career, Mann had personally poured his energy into both a major insane asylum, as they were then known, in Worcester; and, in Boston, the New England Asylum for the Blind (known today as the Perkins School). To Mann's ideological cohort, a widespread system of public schools would be the one institution to rule them all: capable of churning out the sort of Americans who would rise above the general ferment and govern the young United States wisely.

The question Mann and his compatriots faced, then, was how not merely to build a better system of education, but to build the right *sort* of system, which would produce citizens, not partisans; a generation motivated by public spirit, not self-interest. Lancaster's schools, with their students crawling all over one another in a constant scramble for advancement, couldn't have been less suited for the job. No, what was needed was a flat sort of institution, capable of providing children with a common set of experiences. One particularly appealing model could be found in the professionally led classrooms that prevailed at the time in Prussia (part of present-day Germany). As a popular book proselytizing for that model argued in 1836, "The masters of our primary schools must possess intelligence themselves, in order to be able to awaken it in their pupils; otherwise, the state would doubtless prefer the less expensive schools of Bell and Lancaster."

By the late 1840s, monitorial schools in even the Lancasterian redoubts of New York City and Philadelphia had begun supplementing their monitors with apprentice teachers, who eventually replaced them outright. Soon, students at these once-monitorial schools were being taught solely by professionals. And thus the U.S.'s monitorial movement, Kaestle writes, "ended with a fizzle, not a bang."

But it didn't end only in the United States. By the 1850s, virtually everywhere the tide of monitorial education had spread, it receded. In fact, given how widespread and comprehensive the turn away from monitorial education was, its downfall couldn't possibly have been the result of any one source of external pressure. Rather, it makes sense to think of the collapse of the monitorial approach almost as a technological failure: a breakdown somewhere along the line where the internal machinery of the system (mainly student teachers working for free) rubbed up against the demands of the world outside. Even if permitting students to follow individualized trajectories through school was a good idea in the abstract, in the real world of the eighteenth and nineteenth centuries, it simply wasn't practical to set up such a system that was simultaneously cost-effective (in fact, monitorial schools frequently ran less efficiently than advertised), instructionally effective, and—crucially—stable for the long term. The rare monitorial schools lucky enough to have uniquely talented, committed headmasters lasted longer than most. But every engineer knows that when you're designing a system for long-term stability, you can't count on luck. The unavoidable fact was, to enable students to move freely through the curriculum, you needed something on the order of one instructor per student, and the only way to make that happen at a reasonable cost was to have students teaching one another for free. But that hinged on students showing up to teach—by no means guaranteed, especially as competing, *paying* teaching gigs became more available—as well as parents accepting that much of their children's time would be spent on the instruction of their fellows. All told, perhaps what's more surprising is not that the system failed, but that it persisted as long as it did.

ANDREW BELL'S DREAM: A POSTMORTEM

In the United States today, nearly every state has a public school named after Horace Mann—fifty-four in all. Among historical figures who were neither presidents nor founding fathers, only a handful can claim so many schools. The Rev. Dr. Martin Luther King Jr. has seventy-eight. The Marquis de Lafayette, the French general whom Andrew Bell spotted in Virginia, has seventy-nine.

Neither Bell nor Lancaster can claim a single one.

By the turn of the twentieth century, when the first generation of experimental psychologists began to think seriously about how to use the new science of the mind to improve education, the system they'd inherited was a classroom-based, age-graded one. For the foreseeable future, students would pass through such schools on a predetermined schedule, catching as much education as they could along the way, like carousel riders reaching for rings. John Dewey, in his short-lived laboratory school, flirted with softening the temporal edges of this system. The administrative progressives who won the day, however, saw no reason to challenge the self-contained classroom in their quest to standardize education and sort students by their supposed aptitude.

In a different world, perhaps—a world where the name "Andrew Bell" graced school entrances—students might simply have sorted themselves, "graduating" from a given course whenever they demonstrated near mastery of it. But we didn't, and don't, live in such a world. In our world, the vast majority of students, in all their varied, individual glory, are expected to move along at the same rate in every class. Today, we continue to rank them based in great part on their performance in classrooms that may be moving too fast or too slow for them at any given moment. We remain unable to say how they might fare at a more optimal pace.

Even the TEAL classroom, despite the fact that it was designed to meet the cognitive demands of learning minds, suffers from this unfortunate fact. For instance, in the old lecture format, Peter Dourmashkin said, cramming then forgetting was the norm, despite its cognitive drawbacks. TEAL has helped counter that impulse somewhat, but "we still have that problem a little bit," he said. "That problem hasn't gone away"—because TEAL is still part of a larger, time-centric whole.

If an approach as radical as TEAL can't break the spell of time-centrism, perhaps it can't be done—at least, not within the confines of our inherited educational structures. Outside such structures, meanwhile, wild, new things would presumably be possible. But there's no need for conjecture: In fact, a variety of new schools are now springing up that are not only pushing the limits of education as we've inherited it, but leaving them behind entirely. And that's where we're headed next.

- VII -

OUTSIDE IN AND AT SCALE

In the science fiction novel *The Hitchhiker's Guide to the Galaxy*, there exists a simple, unitary solution to "the Ultimate Question of Life, the Universe, and Everything," as its author, Douglas Adams, put it. After mulling over the Ultimate Question for seven and a half million years, an absurdly overpowered supercomputer spit out a perplexingly pithy answer: the number 42. The solution meant nothing to the cosmic scientists who built the computer, but it did expose a fatal flaw in their reasoning. They realized, too late, that they had never defined what the Ultimate Question was, exactly—and it turned out that an answer without a question wasn't much of an answer at all.

To two young humans back on the real Earth, however, 42 represented the answer to a problem that felt just as weighty, albeit far easier to pinpoint, since it followed them around like a personal raincloud. The working world, they understood, demanded credentials of higher education, but the right degree, not to mention the right training, was perpetually out of reach.

For René Ramirez, a bachelor's degree in biology at San Francisco State University had proven elusive. His senior year, he had already been on academic probation when his grandfather died and René took on some breadwinning responsibility for his family. The new stressor pushed his grades over the edge. When the university told him he could pay an extra $3,000 for a remedial course that might or might not set him back on track, he bailed. A follow-up effort at San Francisco's City College proved just as unsuccessful. By his early twenties, he was beginning to wonder if he would ever be able to cobble together a degree.

Josh Trujillo was also scrambling for answers. He'd recently completed an undergraduate program in entrepreneurship at Florida State University, but his first business, a healthcare startup he'd begun in his second year of college, was in a death spiral. The biggest problem, he later said, had been his lack of technical expertise in coding and web design, which his entrepreneurship major had omitted. It seemed like another round of college might be necessary if he ever wanted to launch a successful tech startup, a prospect that sounded both expensive and demoralizing.

Near the end of 2015, both René and Josh were weighing their options when they heard of an opportunity that seemed too good to be true. Xavier Niel, a French telecom billionaire, was opening up a free coding academy just outside Silicon Valley, where even complete neophytes could pick up serious coding chops in the space of a few years for little more than the cost of food. When they checked the fine print, they discovered that there wasn't a catch, exactly, although the school did have a few decidedly untraditional aspects. But these weren't troubling enough to dissuade them. If anything, it seemed like they had traveled as far as they could along the road of traditional education. Perhaps an uncommon path was exactly what they needed.

The most obviously nontraditional aspect of the coding school was its name: the number 42. It was chosen, they found out later, because it was the answer to the Ultimate Question of Life, the Universe, and Everything. It also carried a hidden message: Spoken aloud, the number sounded like "fortitude," which any successful applicant would have in spades.

Unlike most other academic programs, 42's admissions process took into account no test scores, no letters of recommendation, no overwrought essays or overstuffed transcripts. Instead, it accepted all comers into an intensive, twenty-eight-day trial period known as a *piscine*—French for "pool"—which permitted applicants to distinguish themselves solely through performance. Ecole 42, as the organization's flagship location in Paris was known, had already been operating for several years. René and Josh joined the inaugural piscine at 42's first American site, located in Fremont, a city on the edge of California's Silicon Valley.

Today, several years later, 42 Silicon Valley remains a jarring place to walk into. The school is contained within a single blocky building, cloaked in a curtain wall of blackened windows. Earlier, a for-profit university chain had operated a branch out of the same building and, at least in terms of its surroundings, the for-profit had fit in better. For miles in every direction there extends a green-and-black circuit board of lawns, parking lots, access roads, office buildings, and big-box stores: an integrated system of economic, geographical, technological, and cultural forces aligned to promote driving, spending, and earning. When 42 moved in, however, it arrived with its own internal logic so distinct from the world outside that walking across its threshold today feels a little like entering a foreign embassy.

Its unique perspective only starts with the fact that the school, free of charge to students, rejects the idea that higher education is a consumer good to be purchased like a pair of sneakers or a gallon of milk. It also rejects the surrounding area's car culture: Most students sleep in dorms across the parking lot from 42's academic building, but spend the bulk of their waking hours at school, and eat their meals there as well. Add in the fact that the admissions process demands no input from the outside world, and a picture of semi-perfect isolation begins to emerge. Monk-like, 42's students turn inward, toward the blackened edifice they call home.

Which, upon closer scrutiny, is not entirely wrapped in smoked glass. A horseshoe of white concrete that wouldn't look amiss at Stonehenge spans the entrance. Inside the cavernous main workroom, too, rows of empty white doorframes extend in different directions, rising like croquet wickets above the students seated at desktop computers. To 42's student body, they represent the school's *academic* gates, which continue long after admission. 42's entire curriculum is arranged in sequential stages, like a video game, and the only way to advance is to turn in an acceptable coding project. Students proceed thus, stage by stage, along one of several available paths, which are twenty-one stages in length. At the time of this writing, although a few students are getting close, no one has ever finished level 21—not in Silicon Valley or in 42's other twenty-odd global franchises, financed by wealthy individuals and governments around the world. Before they can make it so far, 42's students tend to be drawn away by attractive job offers—a remarkable outcome, given that many enter the program knowing not a single line of code.

Today, 42 Silicon Valley's students have found jobs at almost every major local tech company and many lesser-known startups. Part of this ongoing success may have to do with the fact that the piscine selects from the start for dogged, insightful workers. But there's more to it than that, and it likely has to do with 42's stage-by-stage structure. The scheme is an example of a pedagogical philosophy that, in other schools and contexts, can sometimes produce outstanding results, but which is either difficult to pull off consistently, or else costs so much to run, that, in its hundred-plus-year history, it has rarely been seen in the wild.

In "mastery learning," as the approach is known, the premise is simple: You don't advance to Concept B until you've demonstrated total command over Concept A. Test scores serve not as a means of comparing you against your classmates, but rather as a way to determine whether you're ready to advance. And the bar is set high: in some versions of mastery learning, at a score of 90 percent or more. The modern name was bestowed upon the scheme in 1968 by the educationalist Benjamin Bloom. Similar approaches, however, had found different champions, with different motivations, at different times in history. An ur-version of mastery, for instance, prevailed at Bell's and Lancaster's monitorial schools, which permitted students to advance through curricula at their own pace. Later, in 1919, a superintendent in Cook County, Illinois, created the mastery-based "Winnetka Plan" in response to the then-ascendant systems of grading and advancement. Later still, in the 1950s, one of B. F. Skinner's chief selling points for his teaching machines was that they enabled a mastery-type progression: freeing students from the tyranny of the classroom clock (at the low, low price of chaining their attention to their desk for hours on end).

Today, mastery is experiencing something of a resurgence. More than forty schools in New York City have instituted mastery-based systems as of 2017, and schools in Vermont, Maine, New Hampshire, Illinois, and Idaho are either following suit or testing the approach. Success stories abound: For instance, after instituting mastery learning in 2014, one struggling New York City school saw its percentage of students reading at grade level jump from 7 to 29 percent in two years. But for every result like this, there is the unavoidable rejoinder that virtually *any* change made in overburdened classrooms could be beneficial. And meanwhile, concerned parents rightly wonder if modern mastery learning systems, many of which require students to spend lots of time seated at desktop computers, are little more than updated versions of Skinner's teaching machines, with their purported pedagogical benefits really serving as a stalking horse for their cost-efficiency.

Mastery learning, however, is just one of several possible routes to personalization. Hiring a highly skilled personal tutor, for instance, is one effective, albeit expensive, way to create a personalized learning trajectory. Various outside-in aligned forms of education offer other, similar routes: with close enough oversight on the part of teachers, it's possible for students to learn what they need to know while following their own interests at their own pace. The rub, in both cases, is that they don't scale well. Even if cost were not an issue, both tutors and successful child-centered schools rely on exceptional teachers—which is a problem. I firmly believe that to the extent that human flourishing exists in this world, we mainly have exceptional teachers to thank. But exceptions, by definition, are the sort of thing you can't count on when you're working at a population scale.

Now, however, a small but increasingly influential set of tech-enhanced approaches is expanding the realm of the possible. 42 is just one of several outside-in-leaning programs that are developing technologies in an attempt to deliver broad, deep, activated knowledge in a way that is cognitively user-friendly, personalized, consistent in terms of quality of instruction, and, in theory, scalable.

In 42's case, its scaled-up personalization comes from a hybrid pedagogy: it bolts the sort of project-based approach beloved by discovery learning advocates onto a programmatic curriculum. As its students work their way through, they "level up" not via multiple-choice tests, but rather by figuring out how to build increasingly complex edifices of code.

Because the projects 42's upper-level students complete are similar to the challenges they will encounter in the working world, the approach promises job-ready skills. It was this offer—of not just another academic credential, but rather real coding superpowers—that led Josh and René to uproot their lives and move into 42 Silicon Valley's dorms for its first-ever piscine. What followed was twenty-eight days of red eyes, caffeine, cafeteria food, and unbroken work. Although the piscine was spread over a far longer period than any single, standardized test, the clarity of what the school offered, plus the piscine's cost in time and effort, made the stakes seem somehow bigger—as gargantuan as the concrete gate straddling the building's front entrance. And so, at the end of the piscine, when both Josh and René were denied admission, the blow felt more crushing than either had thought possible.

PLANET CLAIRE

Three hundred fifty miles to the south, in Los Angeles, Claire Wang was seated in the rear corner of her eighth-grade classroom, reading a book by herself while her teacher led the rest of the class through one of its daily lessons. She already knew most of what the teacher was going to say—not the exact wording but the gist of it. Claire had already read each of her textbooks cover to cover, twice, in the first two weeks of school. If she wanted, she could have aced her final exams there and then. And so her teacher let her spend class time paging through books of her choice: Ron Chernow's biography of Alexander Hamilton, say, or Michael Chabon's novels. That same year, she'd taken the SAT twice. She'd gotten a question wrong each time, but if you combined her best quantitative and verbal results, she had a perfect overall score. It was around this time that—bored with traditional academics, despite the best efforts of her teachers to accommodate what were, it was becoming clear, dizzying intellectual powers—she'd begun to get serious about exploring the limits of her brain.

Two years earlier, she'd starred on the Lifetime TV show *Child Genius*, a hybrid reality/quiz show where twelve young students answered difficult questions in a battle for a scholarship. They also were asked to perform feats of memory, and Claire, who finished third overall, discovered that she could memorize the order of a full deck of cards fairly quickly—something many untrained adults can't do, even with unlimited time. Curious about her own abilities, she started frequenting online forums where memory athletes discuss tactics.

Memory athletes, she discovered, often don't have any major innate advantages in terms of working memory capacity. What sets them apart, rather, is how they *use* their working memory: less as temporary storage dump and more as a loading depot where they can efficiently match new information with deeper representations—usually, imagined *places*. Using the method of "loci," as the technique is known, athletes tasked with memorizing a deck of cards will mentally place a given card (say, the jack of hearts itself, or else a visual mnemonic representing that card, such as Jack from the movie *Titanic*, clutching a human heart) at a location well known to them (such as their childhood bedroom's doorway). The key to memorizing and recalling such data points in sequence is to string several such mnemonics along an unchanging route through one's house or neighborhood, and then to mentally walk that route. In such fashion, for instance, one memory athlete has memorized pi to more than 65,000 digits.

With such strategies under her belt, Claire soon fought her way to the top tier of competitive memorizers. The summer after eighth grade found her at MIT, competing in the USA Memory Championship.

Each of the day's trials would cull a few of the thirteen assembled athletes from competition. They hastily memorized and recalled 300 words in order (three athletes eliminated), then 40 bits of personal information about four strangers (three more eliminated), and then were quizzed about huge quantities of information gleaned from sources like the periodic table and the Rock and Roll Hall of Fame, which they had studied in the previous months (four eliminated).

By the final competition—memorizing a double deck of shuffled cards in five minutes—only three athletes remained. One of them was the thirteen-year-old from Los Angeles.

She finished third overall. When the competition ended, local journalists descended, notebooks in hand. A few talked to the champion, but most milled around Claire, whose youth made for irresistible human-interest value.

The next morning, eating breakfast with her family at their hotel,

Claire laughed about the whole thing. She'd begun branching out into other forms of extracurricular academics, including online courses and memory competitions, because school had become stultifying. "Everyone has to learn the exact same thing," she said, and it doesn't matter "if you already know algebra—you have to learn algebra for two more years."

But contemplating her next school year, her spirits remained high.

"I'm going to Ad Astra," she said.

As in: the secretive school co-founded by tech billionaire Elon Musk, housed at the SpaceX headquarters in Hawthorne, California.

—

To get to Ad Astra, heading south, you take a right at the structure that looks at first glance like a grain silo, and then drive for the better part of a mile, keeping the row of white oil tanks on your left. The silo, on closer scrutiny, is actually the first stage of SpaceX's Falcon 9 rocket, a 156-foot-tall booster that famously touched down as gently as a falling leaf for a first-of-its-kind vertical landing in 2015. The seeming oil tanks, meanwhile, are in fact a single, unbroken cylinder used to test the scale model of the Hyperloop, a high-speed train that avoids air resistance by traveling through a vacuum tube. Both ventures, as well as the electric car company Tesla, the brain-computer-interface endeavor Neuralink, and assorted other ventures, are headed by Musk.

Nestled in among the mile of SpaceX buildings, behind two layers of friendly security guards, a towering hedgerow, and a form guaranteeing you won't post any pictures to social media, sits Ad Astra.

There's one other hurdle every visitor must face. The second security guard hands you a sheet of paper labeled "Polis," which features 120 multicolored squares arranged irregularly. Each square represents a neighborhood dominated by one or more political parties. "In a way that is ethically sound," the instructions read, "your task is to draw five electoral districts."

One might assume, given its proximity to Musk and SpaceX, that there is something overwhelmingly high-tech about the school. Not exactly. More than anything else, what sets life at Ad Astra apart is the fact that its forty-five students are constantly embroiled in these sorts of elaborate games: what Ad Astra's co-founder Josh Dahn calls "simulations." Dahn invents them all—he describes the process as "just maniacally producing content"—and they can range in duration from less than an hour (Polis is on the shorter end) to longer than a month. Often, students, working in groups or as individuals, are involved in several at once.

A warm October day several months after the memory championship found Claire and the rest of Ad Astra's students embarking on an eight-week-long simulation named "Moses," after Robert Moses, the "master builder" of New York City, who made twentieth-century urban planning decisions that advantaged wealthier New Yorkers, frequently at the expense of residents of color. Projected in front of a central meeting room was a game map comprising rows of empty hexagons. Ad Astra's sixteen oldest students—none older than fourteen—sat rapt while Dahn explained the complex rules, which involved a convoluted land-bidding process. Claire leaned over to whisper to one of her teammates, her eyes fixed on the map. Finally, when Dahn gave the signal, the students scattered into glass-walled, satellite conference rooms to strategize. Almost immediately, one kid emerged and asked Dahn if it would be okay to spy into other teams' conference rooms.

Dahn considered for a moment. "That's not in the spirit of the game," he said.

"But we can still do it," the student ventured. Dahn raised an eyebrow, and the student turned on his heel and ran back to his group. The question was a good one, because the lines separating game play from real life at Ad Astra aren't always clear. For instance, in addition to their academic schedule, personal projects, and simulations, each student maintains a trove of virtual currency: the Astra. Alone or in small groups, they develop business concepts to earn more. Three times a year, the students hold a bazaar where Astras change hands wildly. The students run the entire enterprise, and their duties include coding the marketplace's digital exchange as well as coping with Dahn's imposed tax system.

His students momentarily occupied, Dahn took a minute to explain. He pointed to a poster for the upcoming event. "Right above us here, this is the Ad Astra lottery, which has become this really controversial piece. Like, should they be allowed to have a lottery?" It's up to the students to decide, although Dahn and the other teachers encourage them to think through any ethical contingencies. In addition to floating contentious gambling schemes, students pay each other to redesign their companies' logos, recode their websites, and manufacture merchandise. One of the more surprising side effects of the Astra has been the development of bequests. "There was never a policy created as to what happens when someone graduates," Dahn said, "so companies have been passed down with different amounts of money."

Once, when Dahn came to MIT to give a talk about game design, he began describing a simulation he'd made where turning a monetary profit would account for 40 percent of students' success. One of the researchers stood up and asked point-blank why he was incentivizing students to turn a profit at all. Dahn was astounded. "Do you really think, in the real world, that profit is less of a motive than 40 percent?" Working the forces of capitalism into his simulations, instead of creating simplified conditions, "just feels more real," he said.

And activated knowledge, leading directly to influence over real-world outcomes, is the ultimate goal of Ad Astra. More than a hundred years earlier, in his Laboratory School, John Dewey had set up a microcosmic society of students and encouraged them to take on adult-like working roles: not to study for the benefit of some hypothetical future self, but rather to strive for success in the present. Such an approach has a few intrinsic benefits: It can encourage the curiosity of youngsters better than external motivators like letter grades and gold stars, especially for those unable or unwilling to delay gratification. Best of all, knowledge learned in such settings comes already contextualized for real-world application.

Through its complex simulations, Ad Astra does something similar, though with two key modifications: It assumes a state of advanced capitalism in the world it models (some may find this cynical, others pragmatic), and it assumes a leadership role in that world for its students. If, at 42, you work your way through projects like a character in a first-person video game, at Ad Astra you often hover high above the game board, assuming the role of politicians, city planners, and business leaders. In the adult world, "people who have done great things can really be awful people," said Dahn. The school's mission is not just "to prevent that, but to actively work in the other direction"—to churn out future leaders, yes, but ones who are "imbued with some sense of morality, thoughtfulness, and able to take respect of others into account."

Put so starkly, it may sound like the two approaches—life on the game board versus life above it—stand in opposition, but that's not quite right. Rather, they represent two prongs of a single outside-in push to stem the wastage of human potential perpetrated by standardized education as we know it. In the same way that you can increase the value of a fraction by taking from the denominator or adding to the numerator, educators can expand access to learning by recapturing students unfairly deemed unfit by the educational winnower, while at the same time removing the unnecessary fetters imposed on even the most obviously talented students, such as Claire.

Which isn't to say that every one of Ad Astra's students would have survived the traditional winnower. "One of our most remarkable students, who got a full-ride scholarship to the top all-girls high school in the city," said Dahn, had been overlooked for years. Only when she nearly won a national scholarship for "gifted" kids did her teachers and parents realize her hidden abilities. After she performed so well, her mother wondered aloud to Dahn, "I don't even know what giftedness means. What does that mean? Is my daughter gifted?" He smiled sadly. "It's like, your daughter is, by any measure, exceptionally gifted in a traditional sense, and beyond that, she's an amazing, remarkable human."

Failing to invest in someone with unmistakable capabilities is one of the most head-slapping errors a society can make. Such students, said Dahn, are "a natural resource"—people we need, now more than ever, to work on the many problems encroaching on humankind. The question that arises with these sorts of eager young scholars, then, is how to avoid tripping them up as they construct their personal tree of knowledge. They have demonstrated the ability to absorb great quantities of challenging information, presented in virtually any format. (In fact, in Claire's case, that ability includes a knack and willingness to memorize *reams* of unstructured, encyclopedic data.) In such rare but powerful cases, the enemies of learning begin to look less like the standard nuts-and-bolts challenges posed by forgetfulness, confusion, and mind-wandering, and more like issues that arise higher in the cognitive high-rise: boredom, burnout, and even, if things go wrong for too long, resentfulness of the very experience of school.

Today, Claire seems to be clear of these pitfalls. Seated at a worktable near Ad Astra's outdoor basketball court, she was ebullient. "Well, right now, I'm taking, like, hard classes. I'm taking calculus," she said. Specifically BC calculus, a course completed by fewer than five hundred Americans of Claire's age each year. In most courses at Ad Astra, students pick the overarching direction of their largely project-based inquiry, and then teachers work with each student to set appropriately challenging goals. As a result, on any given day you might see fourteen-year-olds collaborating with seven-year-olds, diving into the same topics at different depths. But calculus is a little different: only three of the school's oldest students take the course, and unlike every other class, it comes with a textbook. Claire said that, like before, she could still plow through the textbook in a couple of weeks, but she no longer wanted to, because her new class lingered on tricky concepts and applications of the mathematics that strayed far beyond the book's pages. She laughed. "I think one of my favorite classes here is calculus. I used to think math is boring but now our teacher's amazing. And it's just the three of us."

For the first time in a long time, school is proceeding at the right pace for her: fast enough to feed her curiosity, but not beyond her (prodigious) limits. This level of personalization is one important part of why, when student-centered, outside-in education works, it can work extremely well.

But Ad Astra is unique: uniquely resourced, with a unique student body, and, most important, granted uniquely free rein to try new tactics. "It's an anomaly because in almost any other world, I don't probably create my own school," said Dahn. "I would never be given the opportunity. I wouldn't get the benefit of the doubt, of being associated with Elon. I wouldn't have the resources of SpaceX. I would never be able to hire the caliber of people that we have. I'd be struggling to sort of evangelize to a group of parents, who are maybe open to a new model, but skeptical."

When it comes to the task of unleashing learning potential on a truly vast scale, however, the things that make a stand-alone school special can turn into liabilities. Trying to multiply an approach like Ad Astra's—utterly reliant on the vigilant energy of a small cadre of highly skilled teachers—is a recipe for failure, one that leads either to costs that, without a billionaire's munificence, fall outside most communities' reach; or else a classroom experience that doesn't measure up to the original. When you're talking about outside-in schools, with their disavowal of conventional classroom structures, the latter is especially dangerous, because when failure does happen, it tends to happen quietly. Say what you will about standardized tests, but they can give you a good indication of when things in a school have gone horribly wrong.

And in any case, the free hand enjoyed by Dahn and a smattering of other lucky educators around the world is simply not forthcoming at a mass scale. Both 42 and Ad Astra, with their unique customs, rules, and even internal economies, feel like outposts of alien civilizations, precisely because they're designed *not* to jibe with the norms endorsed by figures like Mann and Thorndike. And so, like pieces from an Erector set jammed into a wall of Lego bricks, they stick out—looking out of place while hinting that there's another way of doing things.

That continuing incompatibility with the reigning system remains one of the biggest obstacles preventing such visions from spreading. And yet, to listen to certain progressive educators, the time has never been riper for a takeover. The best pathway for this revolution varies depending on whom you talk to: some hope to infiltrate and convert old institutions; others plan to sidestep them; and still others prefer to operate as though the old rules never existed in the first place. Regardless, a single thread unifies all of these efforts. Educational technology, heretofore mainly the province of inside-out, programmatic assaults on learning going back to Skinner's teaching machines, has come to holistic, outside-in learning in a major way.

ON THE OUTSIDE, LOOKING IN

AltSchool, as the Silicon Valley company Altitude Learning was known between 2013 and 2019, took the idea of outside-in research almost absurdly literally. High above its classrooms, AltSchool's ceiling-mounted video cameras and microphones recorded silently as children went about their school days. Machine learning algorithms then trawled the raw video and audio for patterns, in an attempt to identify classroom practices that could optimize learning.

Although I have qualms about some of its methods—recording students chief among them—AltSchool's larger project was potentially revolutionary. At the beginning of the twentieth century, when a standardized, tracked vision of school won the day, it succeeded in great part because it was the most measurable approach. Comparing Thorndike's vision of school against Dewey's wasn't like comparing apples and oranges so much as the *number* of oranges against the *color* orange. To an increasingly efficiency-minded culture, only the countable method made sense.

Today's world is, if anything, more quantitative, a fact AltSchool made no attempt to change. Instead, it posed a simple question: *What if we found a way to count the uncountable?*

In October 2018, AltSchool's founder, Max Ventilla, bounded into a conference room in his company's brick-walled headquarters in San Francisco's SoMa neighborhood. Lanky, bearded, and clad in a plain T-shirt, he exuded Zuckerbergian tech aesthetics. Mark Zuckerberg was, in fact, one of the first investors in AltSchool, which began as a system of centrally run schools, but soon pivoted—to borrow the local terminology—to a different structure. By 2018, the organization was operating just a handful of "laboratory" schools. These informed the development of the broader AltSchool tech platform, which could be purchased by any school seeking to lend structure and accountability to the freewheeling business of personalized, student-centered education.

Ventilla got his start in Silicon Valley at Google. He left to found the startup Aardvark, rejoined when Google acquired Aardvark, and then rose to serve as the head of personalization across Google's products. Today, he's one of a small set of ex-Googlers who say they are taking deliberate aim at the sorts of difficult problems that Silicon Valley has yet to even dent. In a sense, the internet has long been like 42: not the coding academy, but the answer without a question from *The Hitchhiker's Guide*. When Silicon Valley companies, armed with a solution in the form of internet-connected technologies, began to search for problems to solve, they raced to claim the easy ones first. Did we necessarily need dog food delivered by mail, or unceasing updates about the lives of semi-forgotten acquaintances? No, but such non-concerns were the most solvable, so they got solved first. As Seth Sternberg, a former colleague of Ventilla's at Google, has said, most Silicon Valley products have started from the position of "building something that's easy to build, and you don't know if people will want it."

Within the subfield of education technology, Ventilla argued that the same exact process had been taking place for decades. "I think the way that we've used technology in an education context has historically been very dangerous," he said. "It's been to simplify the problem to what technology can solve."

According to outside-in-inclined educators, there's far more involved in successful learning than can be easily automated. Ventilla walked up to a whiteboard and began drawing what looked like a snail's shell. "If I go back to Dewey," he said, "he's kind of describing this experiential learning cycle."

In his 1910 book *How We Think*, Dewey broke down what he called "a complete act of thought" into a five-step process: identifying an open question (what he called a "felt difficulty"), getting a sense of what might be causing it, coming up with a possible solution, testing that solution (through actual or thought experiments), and then either arriving at a conclusion or probing further. According to this theory—and others like it; several educators appear to have crossed this same bridge independently—you undergo this sort of process every time you personally figure out something new about the world. Today, open the pedagogical hood of many outside-in-leaning schools and you'll see the same spiral shape powering the enterprise. Even traditional schools teach it in the form of the famous, five-step "scientific method":* *identify a question, form a hypothesis, test said hypothesis, analyze your data, and come up with a conclusion.*

To those who, like Ventilla, have accepted the helix of experiential learning into their life, the problem with most instructional technologies (with exceptions like Scratch and Logo) is that they skip steps. Going all the way back to the mechanical algorithms

* A "method," it's worth mentioning, that professional scientists often eschew.

of Skinner's bronze-and-wood teaching machines, the educational software industry has poured most of its energy into the subsection of the learning cycle that is easiest to automate: the identification of a problem and suggestion of a solution.

"What ends up happening is you get stuck in little parts of the cycle where you're engaging in the learning very linearly and superficially," Ventilla said. Knowledge gained in this way becomes yet another solution in search of a real problem: like the number 42 in *The Hitchhiker's Guide;* like the offerings of first-wave dot-com companies. Unmoored to any "felt difficulty" that the student actually cares about, it remains as inert as helium gas.

Through its tech platform, AltSchool hoped to support the full learning cycle. Every morning, teachers printed out a daily "playlist" of activities for each student to follow, based primarily on her individual interests and progress, while also making sure that she didn't miss any important topics along the way. Most of the time, students would interact with printouts, not screens, "but on the backend there's an actual app that the teachers are creating those playlists in," said Ventilla. As each student moved through her playlist, she and her teacher recorded her progress in AltSchool's software.

"I have a seven-year-old daughter who is now in her third year of one of our lab schools," Ventilla explained. "She led the last parent-teacher conference that we had." It was like a typical parent-teacher conference—thirty minutes, conducted in the classroom—"except my daughter is sitting there between my wife and I and across from her two teachers. She's got a template that she's filled out with her teachers that has nine different things that she's going to go through.

"She goes through what are her goals. Some of those are academic, some of those are non-academic. They're set and displayed in an app that we built. She goes through what progress she's made—and that is a mix of her reflection, teachers' evaluation, and then standardized grading."

That last part—the standardized grading—may come as a surprise, but to Ventilla, it was essential. "We take a nationally benchmarked, value-based assessment, which shows how much progress a kid has made in core subjects like reading, English, and math," he said. "It's one of the things I think a lot of progressive schools get wrong, is they say: 'You want kids to develop a sense of self, you want kids to enjoy school, you want kids to be connected to each other and their community.' But if that kid thinks that two plus two equals banana," he said, his eyes opening wide, "they're not necessarily going to end up being well suited for the future."

At that point in time, in addition to AltSchool's three laboratory schools, its platform undergirded one public school system and six private schools. The laboratory schools stopped at grade eight, although some of the schools using the AltSchool app ran all the way up through grade twelve. Graduates' prognoses were good: "Over years and years and years, we've had every one of our graduates get into a first- or second-choice school," Ventilla said. Selection effects aside, part of the reason may have been the sheer depth of the transcript compiled by AltSchool's app, which included not the sorts of number and letter grades that merely stand in for learning, but the actual residue of a student's educational journey: a portfolio of projects. "They've got this layered representation of who they are and what they've done. It lets a school make a much more confident prediction that that's a kid who will do well in their school."

—

In June 2019, AltSchool pivoted yet again, rebranding as Altitude Learning. Declaring its research and development phase complete, it handed away the reins of its laboratory schools and began to focus solely on its software. The shift raised eyebrows in Silicon Valley—"Zuckerberg-backed startup that tried to rethink education calls it quits," declared the *San Francisco Chronicle*—and further afield as well. AltSchool's swerve provided ammunition to critics who had wondered if its supposedly holistic approach had merely concealed the same old reductive thinking, indistinguishable from that of earlier generations of tech-toting personalizers. Perhaps, as one disillusioned former AltSchool teacher suggested on his personal blog, AltSchool's "playlist"-based system had sacrificed important aspects of the student-teacher relationship. "It was disembodied and disconnected," he wrote, "with a computer constantly being a mediator between my students and me." In this line of criticism, instead of solving the truly "hard problem" of scaling up and regularizing top-notch, student-centered education, AltSchool had merely blundered its way into addressing yet another easy-to-solve *part* of the problem.

Tidings of Altitude Learning's demise are decidedly premature. But at the very least, its travails illustrate the challenges of shoehorning an outside-in ethos into institutions built on a more traditional model. On top of everything else, Altitude remains a for-profit company, and it's still unclear whether it will convince more than a few wealthy schools to pay for its platform. And so even if it somehow were to allay the concerns of its staunchest critics, it remains to be seen whether it can win enough converts to exert significant influence at vast scales.

Sometimes, however, the fastest way past an obstacle isn't through, but around. Step off the street and into a Wildflower Montessori school, and you'll behold an organization with aspirations every bit as large as Altitude's, but with no intention to foist its approach on existing institutions.

—

The first thing you'll notice when you walk in is how you lower your voice to match the quietness of the twenty-odd four-, five-, and six-year-olds in attendance. They aren't silent—a low hum of activity and conversation prevails—but there is little shrieking or crying, no running or fighting. Instead, you'll notice the young students seated at tables and on the floor, concentrating on the materials in front of them. They're so focused, in fact, that the overall experience feels like sitting in at someone else's place of worship. You're careful not to let your presence become an intrusion.

About half the students sit by themselves, and half are gathered in groups of two or three. One kid is drawing with a purple crayon clenched in his fist, and will continue to do so, imperturbable, for at least fifteen minutes. Another arranges colorful sticks of different lengths into a triangular wooden frame. Another boy, who can't be older than four, lays plastic cut-out letters onto a floor mat, next to a column of pictures. A teacher—there are two full-time teachers and a parent volunteer moving quietly through the room—kneels down and whispers instructions to him. Later, when he puts away the mat and the cut-outs and picks up a pencil to start a different activity, there's something strange about the scene that you can't quite pinpoint. Then it hits you: Never have you seen someone so young holding a pencil so perfectly.

At Wildflower and other "high-fidelity" Montessori schools—that is, those that hew closely to the original template laid out by the movement's founder, Maria Montessori—students spend two three-hour chunks of their day playing like this. Or perhaps they're hard at work. Or both. To a child constructing an abstracted model of the world in the glorious, complex, ever-shifting latticework of her brain, the line separating play and learning isn't so clear.

Nor, according to the now-hundred-plus-year tradition of Montessori education, should it be. Maria Montessori's life's work mirrored—and in some ways presaged—that of Jean Piaget, who conducted observations at a Montessori school for his seminal 1923 book *The Language and Thought of the Child*, and later served as president of the Swiss Montessori Society. Both conceived of learners as active agents running constant information-gathering routines, not blank slates to be filled with facts, and they even both went so far as to describe a sequence of developmental stages, which they thought occurred naturally as children accrued knowledge. Looking back with the advantage of perfect hindsight, perhaps the most important difference between Piaget and Montessori was not of message but of medium: He poured his research findings into the academic literature, while Montessori put her observations, collected over the course of decades, into practice within the walls of the schools that bear her name.

Montessori grew up in Rome, Italy, and, after years studying the natural sciences, attained a medical doctorate in 1896—an astonishingly rare achievement for a woman at that place and time. In the same short stretch of years that, in America, saw the birth and decline of Dewey's Laboratory School and Thorndike's rise to prominence, Montessori began working with children who had been diagnosed with forms of mental disability. In Rome, this frequently meant institutionalization, even solitary confinement, and inhumane treatment at the hands of attendants, who would carelessly throw food into children's barren rooms at mealtimes. As the education researcher Angeline Lillard has written in her excellent book *Montessori: The Science Behind the Genius*, "Montessori saw in their grasping at crumbs of food on the floor as starvation not for food, but for stimulation." To aid them in that quest, she introduced a set of physical, wooden stimuli and began teaching her new students how to use them. In 1901, they passed a national education test intended for students with unimpaired intelligence, a result that earned Montessori wide renown.

She responded less with pride than consternation; the results meant most students were capable of so much more. Perhaps, however, the methods she'd introduced in an institutional setting could benefit children everywhere. Over the next fifty years, while her schools spread around the world, she continued to develop, field-test, and modify her materials and approaches. "Generalizing her discoveries with unparalleled mastery," Piaget later wrote—he understood as well as anyone how science, like learning, involves the abstraction of general rules from noisy data—Montessori "applied to normal children what she had learned from backward ones." The result was "a general method whose repercussions throughout the entire world have been incalculable."

In students' early years, this method involves extensive practice with toys that are more than toys, which, at Wildflower's one-room shopfront schools, fill shelves on all four walls. These materials serve any number of discovery-learning purposes: for instance, enabling students to ascertain essential truths about quantity, dimension, and the conservation of mass. Although much of the school day is unstructured, the learning environment is anything but. The materials are shelved according to ease of understanding, from left to right, bottom to top, and students let their interest guide which materials they work with. There is just one overarching rule: You're not allowed to use a given object in the classroom until a teacher or older student has shown you how.

Historically, due to the socioeconomic differences between private and public school students, it's been hard to evaluate Montessori student outcomes in a randomized, controlled trial. Nevertheless, in recent years, the evidence has mounted in Montessori's corner. In 1997, the city of Milwaukee, Wisconsin, randomly assigned a group of five-year-olds to either high-fidelity Montessori schools or traditional public schools. The Montessori students outperformed the traditional students across a battery of standardized academic tests and behavioral measures—results that were reinforced in 2017, when a similar natural experiment presented itself in Connecticut. Perhaps more important, in the 2017 study, an achievement gap between lower- and upper-income preschoolers closed in the Montessori group, but not in the control.

But the standout result that high-fidelity Montessori preschool delivers time and again is in the domain of reading and writing. Anyone familiar with Montessori only by reputation might mistakenly assume that the system, given its prioritization of student choice and discovery, would fall into the cognitive disaster that is Whole Language reading instruction. But in fact, although Montessori literacy training does rely on discovery learning of a sort, the overall approach is actually highly phonetic, drawing crucial perceptual connections between the appearance and sounds of letters and word chunks. The process begins well before students even know their letters, with certain Montessori materials, such as wooden cylinders that students manipulate using a small knob on top, that train children's hands for the complex, physically demanding task of holding a pencil properly. (And suddenly, the sight of the four-year-old holding his pencil with the confidence of a much older student begins to make sense.) Materials further along in the writing sequence develop students' ability to draw strong pencil lines, and then, when they are ready, they begin drawing (or is it writing?) shapes that adults would recognize as letters. With a teacher kneeling next to them, they learn as they scrawl each letter-shape to recite certain sounds, which correspond not to the identity of the letter (*aitch*) but to its phonetic sound (*ha*). Eventually, students line up groups of the letters they've been writing and reciting. With the glow of comprehension in their eyes, they marvel as "kah-ah-tah" resolves into "cat." All told, Montessori students discover their way to reading bit by bit, breaking into manageable steps a leap that many find too wide to make all at once. The fact that this process meshes well with everything we know about how reading works in the brain is a testament to the power of holistic, top-down research. Many decades in advance, Montessori's observations predicted findings that would require years of study with fMRI scanners to confirm.

Not all Montessori programs are created equal, however. The Montessori schools that yield the best results appear to be "high-fidelity" or "authentic Montessori" programs—and therein lies a major sticking point. The name "Montessori" is untrademarkable, and the degree to which schools stick to Maria Montessori's time-honed methods varies wildly. Consequently, even locales apparently chock-full of "alternative" schools can be thin on access to Maria Montessori's methods.

It was precisely such a shortage in Cambridge, Massachusetts, that led my friend Sep Kamvar to found the first Wildflower Montessori school, partway between MIT and the Cambridge Public Library. Kamvar—yet another ex-Googler involved with search personalization (and yes, you are sensing a trend)—was working at MIT, not far from Mitch Resnick, when he needed a preschool for his two-year-old. He delved into the early childhood education literature, ultimately alighting on a high-fidelity Montessori approach. But slots in the local public Montessori school were in impossibly high demand, and so Kamvar decided to create his own.

Influencing his thinking was not just the writing of, and about, Maria Montessori, but also the architect Christopher Alexander, who coauthored a 1977 book, *A Pattern Language*, that argued that cities work better when given the chance to develop organically, shaped by the needs of the people who live there, as opposed to by city planners in the mold of Robert Moses. In the book, Alexander and his coauthors raised the idea of a multitude of tiny, "shopfront" schools, open to the street, that could simultaneously cut administrative overhead while breaking the architectural spell cast by large schools—which, the authors wrote, carry the "trappings of control."

When prime shopfront space became available in Cambridge, Kamvar moved quickly, establishing the first Wildflower location. The effort was a repudiation not just of standard pedagogical tactics, but also of the idea, going back to Horace Mann and beyond, that learning belonged in big, centralized institutions. Wildflower's commitment to decentralization is also one of the major differences separating it from AltSchool. The organizations share founders with remarkably similar backgrounds, funding sources (the Chan Zuckerberg Initiative, for instance, has been an important source for both), and a great deal of pedagogical philosophy. They've also both experimented with overhead sensors and machine-learning algorithms in a handful of schools functioning as living laboratories. Kamvar even served as an advisor to AltSchool. But where

AltSchool, now Altitude, strives to fit into existing schools, Wildflower is designed to spring up in the gaps *between* them, like dandelions emerging from sidewalk cracks.

Indeed, Wildflower schools spread using the same process that allows plant roots to defy the best-laid plans of city planners everywhere: mitosis. Each Wildflower school contains the seeds of its own reproduction in the form of two Montessori-trained teachers. Whenever demand for seats exceeds capacity, the school simply splits like a dividing cell. One teacher remains, and one goes off to found another school nearby. It's possible to see this biological growth pattern in maps of Wildflower's nationwide distribution. Wherever a founding school once stood alone, a cluster of schools now exists. They can be found, predictably enough, in wealthy coastal regions like Boston, New York, Washington, D.C., and the San Francisco Bay Area, but hubs have also popped up in Minnesota, Indiana, and San Juan, Puerto Rico.

The overall effect is a system of schooling that is, at least in theory, poised for exponential growth, while remaining cost-effective, quality-tested, and equitable.

Wildflower schools are designed to be "actively anti-racist," said Alison Scholes, who helps coordinate the organization's expansion in Massachusetts. "There are a lot of kids in our classrooms who are coming in from low-income backgrounds or backgrounds with trauma, and are able to close that so-called achievement gap in a way that a more traditional preschool, or some of the other early-ed environments, haven't been shown to do." Part of the benefit stems from the fact that an individualized approach is inherently cosmopolitan: a student might choose to practice sound-spelling the word *car*, for instance, but she could just as easily choose *pez*—the Spanish word for fish, swimming in the tank in the back of the classroom. Wildflower is also pushing cultural inclusivity in more active ways: "We're hoping to support an Afrocentric school in Roxbury getting started this fall," said Scholes, sitting in a cafe around the corner from Wildflower's Dandelion preschool in Cambridge. "The art on the walls will look different than maybe the art on the walls in Dandelion, and different from Violeta, which is our Spanish-English bilingual school."

For schools hoping to achieve such inclusivity at scale, leanness is a top priority. As of 2019, Cambridge's public schools run at a cost of $29,000 per year for each student (the state average is $16,000) and local preschools cost between $12,000 and $18,000 per year. By contrast, Wildflower schools in Massachusetts run at a cost of between $12,000 and $15,000 per pupil, despite paying their teachers competitively. Some parents pay that full amount out of pocket, although many benefit from need-blind admissions and financial aid. But even the full ticket price represents an improvement over many traditional preschools and primary schools in the area. A major part of Wildflower's relative leanness has to do with the fact that Montessori schools have always had a higher student-to-teacher ratio than traditional schools, since students with free rein to follow their interests require less overt governance. Wildflower's small facilities are also easier to maintain than a traditional school's, and its testing costs are nonexistent. And, crucially, it lacks the administrative personnel required to run a school of several hundred or thousand students.

With that lack of administrative superstructure comes the risk—present in all schools but especially dangerous in outside-in programs where student progress is difficult to quantify—of wide variation in quality of instruction. Wildflower flattens out inconsistency in part by forming oversight coalitions within its hubs, and insisting that each individual school's board include a teacher from a neighboring school. But qualitative measurements, too, can provide a perfectly strong measure of whether a given child is doing well. Montessori teachers have always kept exceedingly detailed notes of each child's progress: a paper version, essentially, of Altitude's electronic record-keeping software. Through such records, it's possible to see whether children are meeting important benchmarks.

And then there's the surveillance technology. Wildflower, like pre-2019 AltSchool, has outfitted a few of its schoolrooms with overhead recording equipment, which, combined with tiny Bluetooth trackers in students' footwear and the Montessori classroom objects, constantly feed data to remote computer vision and machine-learning software. Eventually, Wildflower's research team hopes to create an objective, quantified data set that will mesh with teachers' detailed record-keeping, adding to the sum of what Montessorians know about learning in their schools.

This approach, put simply, poses stunningly difficult engineering problems. It's hard to say when, if ever, machine-parsed data of this sort will provide insight that human record-keeping cannot, or if it's even worth the risk that supposedly "anonymous" student data might somehow be misused. Regardless of whether this particular vein of research ever turns up pedagogical gold, however, in a broader sense Altitude and Wildflower are still both on to something: scaling up a form of personalized education that strives to keep teachers at the center of the student's intellectual life. As Ted Quinn, Wildflower's head of research and innovation, put it, personalization-minded technologists in the inside-out tradition are usually trying, one way or another, to "automate the teaching process. That's definitely not the bet we're making. We're keeping the human in the loop."

If you pull back and squint at the larger conversation surrounding personalized education, what Altitude and Wildflower have done is defined a tradeoff. If, on one hand, you want to see personalization at scale, but don't put much stock in the experiential learning cycle, then you can probably build a self-paced, automated system for information delivery. But if, like Dewey, Montessori, and Piaget, you believe that the learning process is actively driven by the student's questing mind—an object of astounding complexity—then it's necessary to meet that mind with an object of equivalent complexity. At present, only another human mind fits that bill. Only concerted human effort can improve on what, say, a textbook alone can do. And now, for the first time, we are glimpsing the rise of teacher-powered, scalable education systems designed to do exactly that. Systems that might someday, at least in theory, re-create Claire Wang's calculus breakthrough, millions of times over.

—

And, strangely enough, we're also witnessing the rise of an experiential-learning approach that defies all the rules. According to the tradeoff defined by Altitude and Wildflower, 42, lying low in its Silicon Valley office park, shouldn't work. It's built on a project-based, discovery-learning model, and yet it has almost no teachers to speak of.

When Josh and René were denied admission to 42, it came as a crushing blow. In the month they'd spent on campus in the piscine, they had glimpsed a tantalizing picture of themselves, wielding control over the invisible realm of ones and zeroes underlying so much of the visible world. But even if either was upset, there was literally no else one to blame. As Gaetan Juvin, the head of pedagogy at 42 Silicon Valley, later explained, 42 was designed from the start "to automatize everything." It has no admissions department. There is no dedicated tech support for the hundreds of computers and twenty-four-hour servers running in each location. There are no textbooks. And, at least in the way that most people think about them today, there are almost no teachers.

Or perhaps *everyone* is a teacher. As 42's students work their way through their coding projects, they turn to each other both for help and assessment. If the scheme seems familiar, it should: It is as close as anything I've seen to the second coming of monitorial schools. Although 42's pedagogues hadn't set out to copy the work of Andrew Bell, a similar set of founding conditions—including a munificent benefactor whose resources were large but not unlimited—led to a similar solution. There is one key difference, however: 42's students teach each other not the facts inside textbooks, but rather how to solve problems. When a student in the midst of a coding project encounters a "felt difficulty," as Dewey put it, she looks for the answer in the project's accompanying materials, or Googles for it. If that fails, she asks a neighbor. The thornier the problem, the more students begin to gather around, surrounding the irritant like nacre in an oyster's shell. At a certain point, someone grabs an advanced student to weigh in. Only the most intractable obstacles make it all the way up to a teacher—but this hardly ever happens. Of the twenty staff members that support the thousand students at 42 Silicon Valley, only four are teachers.

Nevertheless, the system appears to run stably. Its secret is closely related to why 42 is sealed off so tightly from the outside world. It's become common to think of the relationship between student and school in economic terms: The student is a consumer of education, sold by the school. But selling knowledge is different than selling other things. "If I sell you a bottle of water," Juvin explained, "then I don't have the bottle anymore." But when he sells you a fact or a hint to a problem, he retains a copy for himself. Consequently, in a mostly closed system like 42, when students help each other solve problems, the local repository of institutional knowledge only grows.

A similar, albeit more formalized, economy prevails over assessment: If you want a more advanced student to grade your latest project, you must pay them a chit, which you can only earn by turning around and grading the project of someone less advanced than you. This mechanism isn't just cost-effective; it also forces students to revisit past subjects with some regularity, the better for long-term recall. 42's cognitive benefits don't stop there: just the right level of challenge greets students with each new project, a recipe for continued curiosity. And as they teach each other, they exercise neural pathways involved not only in the encoding of information, but also in its retrieval.

All told, virtually all of 42's idiosyncratic rules and customs correspond to benefits found somewhere in the cognitive high-rise. These customs—not to mention 42's ladder-like system for advancement—can make it feel almost like a benevolent cult. And like any cult, once you get involved it can be difficult to stay away.

The piscine that Josh and René failed was the school's first. Before launching its third piscine, the school reached out to offer them another chance. The same rules applied—twenty-eight days of hard labor—except there was another wrinkle. This time, there was no spare room for them in the dorms.

They leapt at the chance. To afford a nearby Airbnb, Josh sold his car. René opted instead to sleep in his, which irritated his lower back until Josh took pity and let him move in with him. If anything, they worked *harder* this time, spurred by their knowledge of what it would take to pass. And finally, at the end of twenty-eight days, they earned the right to walk through the first gate.

—

Today, as the receipts from inside-out cognitive science laboratories continue to come in, it's remarkable to watch them vindicate certain outside-in learning strategies. Meanwhile, for the first time, outside-in pedagogy is starting to look tantalizingly scalable. But perhaps the greatest surprise, coming out of 42, is the idea that project-based learning can serve as the *key* to achieving scale.

In some subjects, that is. Coding lends itself well to a project-based structure, since code is ultimately checkable: Either it works or it doesn't. Other topics, however, lend themselves to different approaches, and sometimes, a bare-bones, inside-out strategy can be more effective. Sometimes it may make sense to delegate "some shallow drill-and-kill style of practice" to an automated system, said Ventilla. "It's not that I think that stuff is a total waste."

The dividing line demarcating where it's appropriate to invoke such systems is on the move, however. Recent advances in data analytics and machine learning, combined with scientists' continued forays into the dark regions of cognitive science, are contributing to a new picture: an inside-out route to mass education, running parallel to the new outside-in road staked out by schools like Ad Astra, 42, AltSchool, and Wildflower. Already, a number of deep-pocketed organizations are developing ways to meet each student's unknowable cognitive complexity not with another human mind, but rather with sufficiently advanced algorithms. And, to the delight of some (and consternation of others, justly worried about unintended consequences) it's showing signs of catching on.

- VIII -

TURN IT INSIDE OUT

For anyone who hopes to make learning more effective for far more people, both outside-in and inside-out approaches now offer real possibilities. We've seen how outside-in thinkers, armed with new technologies and organizational structures, are attempting to replicate top-notch learning practices for sizable numbers of students.

The inside-out strategy of taking successful learning, breaking it open, and reverse-engineering it has shown its drawbacks, however. Most disconcertingly, the approach involves a hard-to-assess degree of risk. A hundred-plus years ago, E. L. Thorndike and his allies thought they'd reduced learning down to its constituent parts, leaving no important gear or sprocket unaccounted for. It turned out that they had oversimplified matters considerably, and the education system built on their blueprint proved reductive: a crude, pixelated rendering of the reality of learning. They had jumped the gun.

Despite the wealth of scientific knowledge we've amassed since then, today's inside-out thinkers face essentially the same blind leap: Do we finally know enough about how learning works to build a better system from the ground up? After all, the cognitive science of learning is still replete with questions not just unanswered but unknown. Add in still more open questions concerning how individual brains fare in any number of different contexts—social, economic, nutritional, sleep-deprived, bathed in stress hormones, you name it—and you've got more than enough uncertainty to give all but the most reckless inside-out educators pause.

And yet the inside-out approach has a few intrinsic advantages, too. First, because the outdated educational structures we hope to improve upon were themselves built on an inside-out ethos, it's not unreasonable to assume that they will more readily take to inside-out repairs, like a surgical patient accepting a kidney from a close relative. Second, once you've stripped something down and can explain it from the inside out, it becomes substantially easier to re-create it at scale. A film photograph might be more beautiful than a digital photo, depending on whom you ask. But only a digital camera can strip a complex scene down into numerical values of red, blue, and green, creating an image file that can be reproduced over the internet with no delay, with no diminishment in quality, at essentially no cost.

And finally—and tantalizingly, and controversially—at a certain point in their technological development, inside-out strategies tend to meet, then exceed, outside-in approaches. You can simply *do* more with a smartphone camera than a film camera. A similarly stripped-down approach to inside-out education might very well present pedagogical opportunities we can't yet anticipate.

It was for all of these reasons—but especially the last one, dangling the possibility of as-yet-undreamt-of improvements to education—that when the time came for MIT to make a concerted, institutional bet on just one route to better learning, we shoved our chips toward "inside-out." We wouldn't just become the purveyor of online educational materials, including a new library of full-fledged courses. We would go even further and establish a pipeline from our community of researchers, within the cognitive science disciplines and beyond, into educational practice. We closed our eyes, gritted our teeth, and rolled the dice.

THE MONGOLIA INSTITUTE OF TECHNOLOGY

In the years following that decision, millions of students have benefited, as will many more, I hope, in the years to come. Included in the very first group of these learners was a young high school student living in a high-rise in Mongolia's capital city, Ulaanbaatar.

"I was in the ninth grade and I was really extremely into math and science," Battushig Myanganbayar later explained between bites of shawarma at MIT's student center. "But the high school curriculum in countries like Mongolia and Russia is a bunch of math drills. You will never understand why you're solving certain equations. You don't even have any clue about how it's being used in real life."

Battushig's school had one advantage, however: Enkhmunkh Zurgaanjin, MIT's first-ever Mongolian alumnus. By late 2011, now serving as principal of Battushig's school in Ulaanbaatar, Zurgaanjin received word of his alma mater's foray into something known as MOOCs. The funny-sounding acronym, which stands for Massive Open Online Course, was first coined in 2008, and became a household term—in a nerdy subset of households, anyway—in 2011, when Stanford launched three open online courses.

In December of that year, the arm of MIT concerned with MOOCs, known as MITx, announced its inaugural entry: Course 6.002x, or Circuits and Electronics. Zurgaanjin encouraged his students to take the free course in their spare time and, when he mentioned that it would help explain how an iPhone worked, Battushig's ears perked up. 3G internet had recently arrived in a big way in Mongolia, and by then even nomadic families living on the steppe had internet access. Battushig's mother, a doctor, and his father, a manufacturer of carpets, had brought smartphones into their home, and they stood out as the sort of thing whose secrets no one would ever willingly give away. Along with nineteen other students, Battushig signed up.

Back in Massachusetts, shortly prior to the course's launch, I was hustling across campus when I ran into Anant Agarwal, the professor of 6.002x, who was spearheading MITx's overall effort while also running MIT's famous Computer Science and Artificial Intelligence Laboratory, or CSAIL. We were both sleep-deprived: I'd been commuting to and from Singapore every couple of weeks, and Agarwal had been working into the night for months. I was excited to find out how the MITx effort was shaping up, and asked him how many students he expected to sign up for his course.

"I hope no fewer than five thousand students because that would be embarrassing," he said, laughing, "but no more than ten or twenty thousand, because I'm losing sleep as it is and I might die."

When Course 6.002x launched in May, 155,000 people registered.

Of these, 23,000 stuck around long enough to attempt the first problem set, Battushig among them. He spoke only limited English, and found himself struggling to keep up with Agarwal at times, but whenever Agarwal switched to the more universal language of mathematics, Battushig discovered he could follow "just based on the equation," he said. After learning each topic, in a manner that would have made Andrew Bell proud, he turned around and recorded an explanatory video in Mongolian for his schoolmates.

When roughly 9,000 students passed the midterm, Battushig was one of them. When just 7,000 passed the course, Battushig was one of them as well.

Ultimately, of the 150,000 who initially registered, only 340 received a perfect score. Battushig was one of them, too.

Not long after the results came in, a journalist from the *New York Times Magazine* came calling to interview Battushig for a piece ultimately titled "The Boy Genius of Ulan Bator." And not so long after that, Battushig, now a local celebrity, began filling out university applications.

As Battushig's future began to unfurl before him, so too did the possibilities for MOOCs. This new form of decentralized, self-paced, online learning seemed capable of saving, or even *recalling*, untold numbers of people from the capricious reach of what I'd begun to think of as the educational winnower. As Agarwal later said, MITx has built an eager audience, including "people that are *just not able* to go to college": a coalition of the unduly winnowed who, for any reason—geography, career, family, age, health, prior test scores, or something else—have found college's demands either insurmountable or too costly.

If MOOCs held great potential for individuals, their capacity for releasing latent talent on a population level was unquantifiable. For every Claire Wang, labeled "gifted" at an early age, there must be hundreds, perhaps thousands, of powerful learners gliding along under the radar undetected for any number of reasons. For societies around the world hoping to solve deep, abiding problems, finding and supporting them was every bit as important as optimizing the education of students like Claire. As the evolutionary biologist Stephen Jay Gould wrote in 1979: "I am, somehow, less interested in the weight and convolutions of Einstein's brain than in the near certainty that people of equal talent have lived and died in cotton fields and sweatshops."

I find this idea extremely affecting. In fact, what I find just as motivating as the world's "lost Einsteins," as they're widely known, is the loss of folks who might not be once-in-a-generation geniuses, but would nevertheless benefit immensely from an abiding relationship with learning, and would put their knowledge to worthy use. Assuming, that is, they're not prematurely winnowed, like I so nearly was.

In fact, from the standpoint of solving big, world-sized problems such as climate change, almost more concerning than how I was nearly winnowed were the circumstances of my salvation: not at the hands of kindhearted monks or nuns or educators-qua-educators, but by an *oilfield service company*. That job undeniably changed my life for the better, but at the same time, the urgency of climate change highlighted the need for more diverse sources of education and technical training. And so part of what captivated me and many other observers about MOOCs was the idea that they might open up new avenues to activated learning for people who might, say, want to work on a solar installation someday, and not an oil rig.

As MOOCs' early cheerleaders soon found out, however, reaching the world's learners would require more than merely tossing them a life buoy in the form of online courses—even top-notch, free ones. Complicating matters, meanwhile, were the legions of new entrants in the larger field that had become known as edtech, of which MOOCs constituted only one part. Within a few years, edtech had become volatile, sloshing with venture money and populated with thousands of companies ranging from tiny startups to venerable giants. That was fine, as far as it went: great things do tend to stagger out of primordial stews. Perhaps inevitably, however, a subset of these companies developed instructional systems much farther out on the inside-out end of the spectrum than MOOCs, evincing worrying levels of confidence in their models of the learning brain. History has not been kind to hubris in education, and although public opinion was smiling on edtech for the time being, it was one fiery wreck away from changing its mind.

I remained convinced that a new, highly personalized, inside-out education system was technologically possible. We weren't there yet, however, and now external pressure was beginning to mount.

SANJAY 2.0

MIT's push toward free personalized learning has roots stretching back at least to the late 1990s. First, in 1997, professors Eric Grimson and Tomás Lozano-Pérez created an online version of their introductory computer science course for their students. To my knowledge, this was the first of its kind to interpolate video lectures with exercises—a structure that is now a mainstay of MITx courses. Next came OpenCourseWare (OCW): a wonderful, radical idea that came about at a tumultuous moment in internet history, when, prior to the first dot-com bust, a number of major universities tried to turn a profit by selling their brand-name education online. It was supposed to be "an absolute goldmine," recalled Hal Abelson, a computer science pioneer and one of OCW's founders, and MIT had arrived late to the rush. In an effort to figure out whether other universities' quasi-private model was worth copying, the Institute hired not one but two management consultancies to do the math—an idea, I'd be remiss not to mention, belonging to one Dr. Gitanjali Swamy, then a consultant at one of the firms, who happened to be married to yours truly. Both firms waxed pessimistic, but one raised an intriguing suggestion: Instead of selling educational content online, why not give it away for free?

Most faculty members loved the idea—myself included. The years following my hasty departure from the North Sea oil platform were perhaps less cinematic than those that had come before, but they were a hell of a lot more enjoyable. Once, I'd had to hold my eyelids open with my fingers as I slogged my way through mechanical and electrical engineering textbooks. In the model of the world I'd constructed in my head, each new fact had aligned only haphazardly against its fellows. Learning had been like trying to fill a dumpster with oddly shaped automotive parts. Without a sense of how it all linked up, the pieces' funny angles and sharp corners only jutted and scraped against each other in ways that seemed to take up unnecessary space. Now, however, not only did the pieces interlock in a way that made sense, but that experience made it *easier* to add on to the whole. And so, post-oil-platform, my time spent poring over books—and increasingly, the internet—felt entirely different, and far more pleasant, than my prior studies.

That wasn't all. Earlier, anytime I wanted to add to or amend my constructed pile of engineering knowledge, I'd always had to dredge up various, shallowly learned principles in order to marry new information to the overall structure. This effort had always imposed undue cognitive load on my precious working memory at precisely the moment I needed it most. Thanks in great part to the spacing and interleaving and retrieval practice I'd been subjected to in my industry training, however, fundamental engineering and physics principles now sat at my fingertips. I could wield such facts as though they were part of me: powerful tools to be used in the construction of yet more knowledge.

It was with this new brain—Sanjay 2.0!—that I began to sail through my studies for what felt like the first time since middle school. On the helicopter ride off the platform, I began to plot my return to the academy. I assembled the coursework I'd need to apply for a master's, which—it all seems so quick in retrospect—led to a PhD in mechanical engineering and, ultimately, a job offer in the Department of Mechanical Engineering at MIT.

Along the way, I began to teach and, inevitably, to recognize the challenges of life on the other side of the podium. When you're responsible for just yourself, it's hard enough to overcome learning's malevolent rogues' gallery: the not-quite-right explanations, the misunderstandings, the confusion and forgetfulness and cognitive load limitations. It's quite another, however, to straighten out such snafus in someone *else's* brain. Learning, I now knew from personal experience, didn't have to be a battlefield. Even predating my time in the oil-flecked wilderness, there had been rare moments when learning had come easily: for instance, when I'd had the chance at university to build a snake-shaped robot, the thought of which still makes me unaccountably happy. But such fleeting moments had always felt like the exceptions to the pattern. Most of the time, I'd understood learning to be a ruthless, competitive business, red in tooth and claw and teacher's pen—and, despite my best efforts to prove otherwise, many of my students seemed to agree.

Then, in 1997, I was invited to help teach Course 2.007 with Alexander Slocum, the legendary professor who inherited it from

Woodie Flowers. That year's robotics contest was sports themed, with robots vying to push balls and pucks onto their opponents' side of an intervening divider. (I still remember how that year's winner, the tanklike Fuzz Bumper, belonging to one Timothy Zue, prevailed over significantly more complex opponents.)

Course 2.007, I began to recognize, did something curious. With its project-first structure, it helped students contextualize engineering far more effectively than a traditional lectures-plus-exams setup, because whenever they encountered a "felt difficulty" in the lab, as Dewey put it, they turned to the lecture content for a solution. And meanwhile, as they built their robots over the course of the semester, they could hardly avoid repeatedly revisiting knowledge that they had filed away earlier in the course—benefiting, as a result, from both spaced repetition and retrieval practice.

Perhaps even more important, the hothouse atmosphere of Course 2.007 made the apparatus of the educational winnower feel like an afterthought. For once, students were competing for something grander than a high GPA. In fact, because the semester-end contest had no direct bearing on students' grades, it appeared to have a freeing effect, permitting students to take risks, make mistakes, and learn from those mistakes in a way that's not possible when your permanent transcript is on the line.

Even while 2.007 was drawing knowledge into my students with an almost capillary action, I had to acknowledge that the same thing was still happening to me. One day, I was talking to my colleague and close friend David Brock, a roboticist who worked down the hall from me. For years, a bulky, expensive system of electronic transponders, known as RFID tags, had been used around the world to track shipments of vehicles and other large inventory items. Brock wondered if they could be simplified significantly (and made so power-efficient that they wouldn't need batteries) by stripping down all the data they transmitted to just a single number, and hosting the rest of it on the internet.

My eyes must have opened to the size of teacups, because I realized that if such super-simple RFID chips could be produced, then perhaps they might be produced absurdly cheaply. In that case, they could be used not just to track massive things moving around the world's supply chains—cars, shipping containers, cattle—but small ones as well. I did a bit of approximation, Mahajan-style, and determined that simple, passive RFID chips could be made for less than ten cents apiece (in fact, they would later be manufactured for less than three cents), cheap enough for my purposes. And just like that, the next decade-plus of my professional life dangled before me, assuming I could figure out how to grab it. If I did, I might create something of lasting value for essentially everyone, since it would drive down the retail price of goods, including food, benefiting both suppliers and consumers.

Together with other partners, Brock and I quickly set up a team, called the MIT AUTO-ID lab, devoted to stripped-down RFID technology. One member of the lab, Kevin Ashton, opened our eyes to possibilities posed by super-cheap RFID not just for global supply chains, but also for local inventories. It could be used to keep track of every book in a library, every box in a warehouse, every sweater at The Gap. Such a technology would even make possible a "smart" anti-theft device, now quite common in retail shops, capable of sounding an alarm only for unpaid-for items. (Ashton also famously coined the phrase "the Internet of Things" to describe a world of increasingly interconnected objects.)

The only question was how I would learn everything necessary to make it all work. I busied myself with the task of making RFID fit the needs of different types of users, which necessitated speed-learning across sprawling domains. I hustled to understand not just the electrical engineering and computer science principles underlying RFID, but also the prevailing concerns of semiconductor manufacturing, the worldwide logistics business, and tech startups. We worked with startups to develop the physical tags, and others to make a better, cheaper tag-reader device. We cozied up to corporate giants, who could champion our protocol over others. I personally started up a company to create the open-source, cloud-based software that the new RFID tags would reference. This all required learning on a scale I had never attempted before, and I won't pretend it wasn't difficult. But this time, I knew how to do it. I didn't always know where everything would go on my schematic tree at the start, but I now knew I was capable of making it fit. So I drank coffee, put in my late nights, huddled with colleagues, and, over time, modern RFID came together. Today, it's no exaggeration to say that the world runs on our system. Billions of RFID tags are now used globally: in retail, in supply chains, for highway toll collection, to find lost pets—even to track evidence in police labs.

MITx

In 2003, while we were hustling towards the finish line for our RFID standard, a computer server under Anant Agarwal's desk blinked to life. Running on it was a quirky little program called WebSim: the world's first virtual circuits laboratory. "I was very excited about online learning," he later recalled, "but I said, 'Look, until I can convince myself that I can do online laboratories, this ain't gonna go anywhere.'" So he hacked together a program that allowed his students to tinker with different virtual configurations of conductors, resistors, capacitors, transistors, and the rest. At the time, deployment of this proof-of-concept virtual lab was enough for Agarwal. "I did not take it and put my entire circuits course online. I did not do that. Shame on me!" he said. Then, in 2008, Salman Khan, an MIT alumnus whom Agarwal had taught in the late 1990s, began posting YouTube videos designed to help students through tricky spots in their studies. This effort, which he turned into a nonprofit named Khan Academy, exploded in popularity. In 2011, a new entrant appeared. Stanford University, as I noted earlier, announced that it would offer three open online courses, which would soon spin off into two for-profit MOOC providers: Coursera and Udacity.

Once again, as in the late-nineties dot-com boom, MIT appeared to be a latecomer to a tech-enabled education party. And once again, MIT would respond by giving away the goods for free.

"MITx was launched on December eighteenth, 2011," Agarwal said, rattling off the date like it was his own child's birthday. Earlier that year, MIT's then provost and future president Rafael Reif convened a meeting of the MIT brass to figure out what to do about the potentially titanic changes coming to university education. Reif had long been working to expand MIT's educational footprint, and Stanford's announcements lent the enterprise a new urgency.

At the time, the prevailing plan was to establish campuses in a number of countries, either built from the ground up or else in partnership with local universities. The poster child was the Singapore University of Technology and Design (SUTD), which we had recently launched with the government of Singapore. I was intimately familiar with it, since I had been tapped to lead its instructional design process.

The Singapore project created a rare opportunity to distill MIT's core pedagogical principles, which went all the way back to the beginning. MIT was founded as a more pragmatic, technical alternative to Boston's nearby universities, a legacy that the Institute's founders memorialized in its motto: *Mens et Manus*, or "Mind and Hand." From the start, it prioritized learning by doing, incorporating laboratory instruction at a time when rote learning reigned. As a member of the class of 1868 attested, "The method of teaching was completely new to all of us. We found ourselves bidding goodbye to the old learn-by-heart method . . . We learned from experiment and experience." Building a new university from the ground up in Singapore created the chance to reconnect with these founding principles. For instance, where MIT insisted that every student experience two semesters of core courses across the disciplines before declaring a major, SUTD would have three shared semesters, a "freshmore year" designed to ground students with a firm biological, physical, computational, and humanities foundation. Where MIT made open-ended design courses like 2.007 an option, SUTD made them mandatory across disciplines. SUTD also cut back on lectures significantly. Instead, TEAL-style workshops, labs, and studios convened—a strategy only made possible by the physical environment we were building, which featured far more open and modular rooms and laboratories than the typical American university.

After nearly three years, with SUTD self-sustaining, the time came for me to return to MIT proper, which was weighing its next move in a world that now contained MOOCs. The mood at the 2011 leadership meeting still favored building bricks-and-mortar schools in the vein of SUTD. I still remember the exact moment when Agarwal, by then the head of MIT's main computer science laboratory, changed that. Why, he asked the group, are we helping open small, new campuses in various cities when we could, in his words, "create a virtual MIT for the whole world"?

It's hard to predict how groups will respond to challenging ideas. All I can say is that, by the end of the meeting, it was pretty clear that MIT wasn't going to go for it. And then, within a week, seemingly everyone's mind had changed. It was like the Institute climbed out of the other side of the bed, stretched, and decided it was now time to place a major bet on online learning.

Agarwal began to move quickly. Squinting at the world of online ed in those days was like standing in the predawn darkness while giant, foreboding shapes lurched in the distance. No one could say what tomorrow's sun would reveal. No one knew which model, for instance, Stanford would pursue—for-profit or non-; private partnership or university department; open-source platform or closed. In any case, it was certain to be big, and within a year or two, other, similar entities would appear. In such an inchoate world, the only thing to do, Agarwal believed, was to stride forward and make our stamp.

It was with such an attitude that MIT announced the formation of its own MOOC venture. Given the Institute's recent history of putting course content online for free, the move raised the possibility of a world of MOOCs animated by a similar spirit of openness: free resources available to anyone, perhaps even running on an open-source platform. Because Agarwal wanted to make this statement before the MOOC field finished congealing, however, he had to announce his intentions first and then start building. The plan hit the internet in December 2011, with the first course slated to begin in February 2012.

The ensuing scramble to honor the announcement unfolded on two fronts: designing the platform where MIT's online courses would live, and building its first proof-of-concept course. The former was given the placeholder title MITx on the assumption that something catchier would eventually come up; it didn't. The latter, meanwhile, was originally supposed to be a computer science course, but every computer science professor Agarwal approached blanched at the timetable.

Leaning back and rubbing his eyes with the heels of his hands, he chuckled. Instead of a computer science course as planned, MITx's first course would be an introductory circuits course, taught by Agarwal himself—a job he would shoulder in addition to developing the platform, which itself sat on top of his nominal duties running CSAIL, MIT's largest research laboratory.

This circuits course, he already knew, would feature something like his own long-neglected virtual circuits lab from 2003, now held together by little more than digital duct tape. He needed a new one to accommodate MITx's likely scale, and that was far from the only thing that needed building. Happily—in his first act of self-preservation since the start of the project—Agarwal gathered an academic murderers' row of helpers.

"Each of them is like the *world's expert* in what they do," he said. "Gerry Sussman"—a celebrated MIT computer scientist and artificial intelligence pioneer—"did all the problem sets," he said, producing "amazing, amazing problems." Chris Terman, a senior lecturer at CSAIL, built the new version of Agarwal's virtual circuits lab, and Jacob White, another professor, built the equation solvers that powered the circuit simulator. "He can make differential equations *sing*," said Agarwal. Meanwhile, Ike Chuang—a pioneer of nuclear magnetic resonance quantum computing who would later help erect MIT's overarching Office of Digital Learning—built another needed set of equation solvers. "Ike Chuang—the guy would go away, and the next day he had a linear equation solver worked out for the platform. Just like that! So, the circuits course, even to this day, is one of the most advanced online courses available."

One thing the course needed that Agarwal had not originally considered necessary was a brand-new complement of lecture videos. He'd planned on saving time by resurrecting older videos of him professing in the lecture hall, but, he said, "my team convinced me to redo them, and in a weak moment, I gave in." The new videos borrowed Salman Khan's approach: laying down audio recordings while writing and drawing on PowerPoint slides using a tablet. (The team found Khan's work so helpful, in fact, that in some cases, they decided not to try to improve on the original and simply linked to his videos.) Meanwhile, they reserved lecture-hall videos for big blockbuster classroom demonstrations, which, as often as not, involved the use of a chainsaw in front of a live audience of students. Perhaps the most maniacal move Agarwal made—and in retrospect, the most prescient—was to teach the online version of the course concurrently with the on-campus course, populated with student volunteers. This decision only worsened Agarwal's time constraints, but it conferred certain advantages. For one, it permitted Agarwal to "flip" the on-campus course: Students watched the online lectures prior to class, and then came in to run through exercises with Agarwal and other instructors in person—sessions that were then videoed and provided as supplementary material online. The rationale for such a structure is, as we've observed, quite compelling: Students can adjust the timing and speed of lectures to meet their own preferences, and class time can be spent actively retrieving information from memory and applying it.

Perhaps more important, the in-person class vouched for the realness of the virtual one. Like backing dollar notes with gold, the fact that the on-campus and online versions of the class were coextensive and equally rigorous lent heft to the latter.

As the semester wore on, stories about the mysterious tens of thousands of students on the other end of the internet connection began to roll in. Battushig's name came to Agarwal's attention, and so did that of Amol Bhave, a young man in India who became so enamored with Agarwal's course that he turned around and created an online learning platform of his own. Both would go on to attend MIT and graduate with flying colors.

MITx, meanwhile, was expanding, and fast. Around the same time that Agarwal's circuits course launched, Reif was talking to the leadership at Harvard. In May 2012, the two universities announced the creation of a joint, nonprofit venture: edX. This new endeavor would piggyback on MITx's software platform, and would soon host online courses not just from MIT and Harvard (although MITx would remain its largest tributary), but an ever-expanding list of colleges and universities.

THE CANTABRIGIAN EXPLOSION

The world of open online learning was off to a roaring start, and for a breathless moment, speculation abounded that MOOCs' mere existence would save the world from itself. "Nothing has more potential to unlock a billion more brains to solve the world's biggest problems," wrote *New York Times* columnist Thomas Friedman in early 2013. Another *New York Times* article declared 2012 the "Year of the MOOC," and it seemed like every year thereafter would be filled with more video-based online learning than the year before. And indeed, by the end of 2018, according to the organization Class Central, which acts as a search engine across MOOC providers, over 100 million students had enrolled in 11,400 courses, provided by more than 900 colleges and universities around the world.

Within just a few years, however, expectations began to whipsaw. Enrollment in individual MOOCs, initially an exponential launch ramp, appeared to stall, due in part to the original hype fading and in part to the increasing diversity of available courses, which brought in more students but spread them thinly. Course completion rates, too, failed to increase over time, which vexed some observers. (This never bothered me—leaving a MOOC unfinished is less like dropping out of high school and more like putting down a long nonfiction book halfway through. Regardless of how far you've progressed, you've learned something.) Far more concerning, however, were the students' socioeconomic statistics, which, within a few years, had started to become apparent. Despite the abundance of heartwarming stories like Battushig's, MOOCs seemed to disproportionately benefit relatively affluent populations. Although other MOOC providers kept their enrollment numbers close to the vest, edX made its statistics publicly available. As a series of independent dives into these data revealed, students from wealthier regions accounted for the overwhelming majority of both edX's enrollments (which remained free) and its completion certificates (which, starting in 2015, began to cost between $50 and $300).

Despite these sobering results, edX was still demonstrably helping many learners who, for whatever reason, didn't fit the traditional profile of college or graduate student. In terms of age demographics alone, for instance, edX's enrollment was on the older side, suggesting a user base of mid-career learners. In Agarwal's very first circuits course, half of the students who passed were twenty-six years old or more, including one who was seventy-four years old; 5 percent, meanwhile, were of high school age. It was clear that people outside the traditional collegiate profile were flocking to edX. And yet, the world's poor—the largest group of would-be learners—were not finding their way to the platform. As a result, less than five years after

MOOCs stepped onto the world stage to all the fanfare of a Vegas prizefighter, their boosters—myself included—found ourselves fending off charges that MOOCs were already, as one Udacity official put it in 2017, "dead."

In fact, they're just hitting their stride. As Agarwal pointed out in 2019: "We're still signing up 100,000 students a week. We're signing up more students a week than we've ever done in our history."

And yet, there was one other reason why overeager observers still wanted to prematurely sign MOOCs' death certificate. Even as MOOC providers continued to plunge forward into the foggy future of online learning, stranger entities began to shamble out of the mist.

—

By the mid-2010s, a venture-capital-fueled bonanza was taking place in the field now known as edtech. New companies were approaching the industry from every conceivable angle: content delivery, student finance, testing and assessment, ed research, classroom management, and many others. In 2018, when one capital firm attempted to describe the entire landscape, it counted a total of fifteen thousand companies, which it organized into a map so complicated, it could have passed for the brain's white matter.

One nexus of the map, labeled "digital courseware," is growing crowded: filling up with tech organizations claiming to promote learning that is simultaneously personalized and grounded in cognitive science. A kaleidoscopic variety of approaches is already in use at schools, homes, and colleges. A school might run flipped classrooms, for instance, as the Clintondale High School in Clinton, Michigan, has, by supplementing its home-grown lecture videos with those of Khan Academy, while hosting said lectures on Blackboard Learn and Edmodo, a popular learning management system and an educational social network, respectively. A number of International Baccalaureate schools around the world now offer courses taught remotely, with a human at the far end, delivered by the online provider Pamoja Education, which students take during set-aside periods of their week. A similar but temporally looser approach can be seen in the abundance of "virtual schools" that now dot the United States, which permit students to take most or all of their classes from home, conferring with teachers online or over the phone. Schools that value temporal flexibility but prefer to gather their students in one place, meanwhile, can adopt what's known as a "flex" model, letting their students flow through temporally unstructured days, guided only by software. And so on.

In the variety of approaches now on offer, perhaps the starkest delineation runs between programs whose students all proceed through roughly the same content at roughly the same rate (perhaps with some impressionist blurring at the edges), and approaches that deconstruct the shared classroom experience into a Jackson Pollock drip painting of individual trajectories, free along axes of both pacing and content. Until very recently, advances in this second, more far-out category were held back by limitations of the underlying software, which, at its core, was built on if-then statements of the sort found in Choose Your Own Adventure game books: *To venture into the spooky basement, turn to page 10; to call the police instead, turn to page 12.*

The algorithms that increasingly run much of our lives today, however, have become far more complex, a trend to which edtech has by no means been immune. As I mentioned in the prior chapter, outside-in, research-minded educators like Sep Kamvar and Max Ventilla have experimented with machine-learning algorithms as a means to sift through student behavior for useful patterns. They haven't yet uncovered much actionable intelligence, but similar techniques applied to an inside-out framework can theoretically produce recommendations that any educator bold enough—or perhaps hubristic enough—can employ *immediately*.

The most venerable fish in edtech's machine-learning lagoon is IBM, which has built its education offerings around its well-known question-answering system, Watson. In 2011, Watson became a household name by besting two prior champions on the quiz show *Jeopardy!*. The main challenge for the show's human contestants is one of information storage and retrieval. Not so for Watson, which, though disconnected from the internet during the contest, still came preloaded with a massive corpus, including the entirety of Wikipedia. The challenge for the IBM team, rather, was the part that comes easiest to the show's human contestants: making sense of the clues spoken aloud by the show's beloved host, Alex Trebek.

Human beings arrive in this world with brain structures dedicated to the processing of spoken sounds—which, as you may recall, are involved in reading as well as speaking and listening. To use a common turn of phrase, one might say we are language-parsing *machines;* except, historically, we've been far better at it than any computer. In the past few decades, however, as available computing power has increased, machine learning has gone from theoretically possible to practical, and the computers have begun to catch up. The idea behind machine learning actually goes back to the neuroscience theory that proposes that "cells that fire together wire together," as the handy phrase goes, first posited by Donald Hebb* of McGill University. In Hebb's theory, variations in strength among synapses "wiring together" could form the foundation of memory. More or less contemporaneously with Hebb's decree, computer scientists began to think about how the brain's probabilistic synapses and neurons could be represented not in flesh, but in computer code. It's possible to set up a single virtual neuron, for instance, that can be trained to "fire" or not (that is, give a yes-or-no answer) in response to many combined inputs (such as the individual light-sensitive cells in a digital camera's sensor). Just one virtual neuron so designed can be trained to identify whether a written shape is a given letter or not—a feat remarkably similar to what the brain's letterbox appears to do. Artificial neurons' pattern-finding capabilities really start to open up, however, when you stack them in tiers on top of one another, such that the yes-or-no outputs from each layer train the subsequent layer. Today, "deep learning" systems can feature twenty or more layers of neurons. In the most complex of these systems, most of what takes place between input at layer 1 and output at layer 20-something is fully hidden from observers. Indeed, this opacity is so complete, and

* And if his name has faded since you first encountered it in chapter 2, it's worth revisiting why. If you remembered the general idea behind his model of memory, then, according to the Bjorks, perhaps competing names, real or imagined, crowded out Hebb's. Or perhaps you put this book down for a few weeks and, without continued stimulation, the synapses involved in your memory representation of the entire fire-together concept became weak with disuse. Here, happily, is a chance to fire those wired-together synapses back up and reap the benefits of spaced practice. See if Hebb's name lingers longer in your memory this time around!

deep learning systems now so prevalent in our lives, that a group of modern computer science luminaries have called for the formation of a new, unified field named *machine behavior*, to study their effects. In a satisfyingly cyclical turn of events, the field seeks to apply the stimulus-response research paradigm of behaviorism—championed by B. F. Skinner, Thomas Watson, and even E. L. Thorndike—not to people, but to computer algorithms.

It was one such algorithm, trained to abstract semantic patterns from spoken English and match them against a deep corpus of knowledge, that permitted IBM Watson to vanquish the most redoubtable human competitors on *Jeopardy!*. While the public was growing accustomed to its new quiz-show overlord, IBM was already working to slide its question-answering system into every conceivable economic niche, in fields ranging from healthcare to finance to the legal profession to even fantasy football. In 2016, as part of this push, IBM announced Watson Education.

As of this writing, Watson Education has come out with two main products, Watson Classroom and Watson Tutor, which are really Janus faces on the same head, oriented toward teachers and students, respectively. IBM's engineers can feed this core system a corpus of information—say, the content underlying a geology course—and, in theory, it will parcel that content into individual topics.

"You have to train it to a domain," said Alex Kaplan, IBM Watson Education's global leader of strategic deals. "This," he said, making a box with his hands, "is everything that's known about geology. So Watson knows all of that. Since it knows all of that, Watson can draw a concept map of the interrelationship of all the primary concepts. So now you've got this concept map around geology."

By a few different means—in the case of Classroom, mainly a classroom's worth of standardized test data; in the case of Tutor, mainly questions posed by a text-based chatbot—the core system then attempts to map a given student's personal trajectory through a course's worth of material.

Tutor, which IBM has developed with the publishing giant Pearson, is supposed to function like a wise owl, sitting on your shoulder and answering your questions as you read through an e-textbook. Once you fire it up, it's designed to zealously pursue a single goal: to get you to demonstrate mastery of each important concept in the unit or chapter, nudging you along with well-timed questions and hints from relevant passages. (For now, only higher education and professional students have access to it.)

This version of mastery learning, different from most classroom mastery schemes, juliennes topics wafer thin. Instead of defining "mastery" as, say, a 95-percent-or-greater score on a course, or even a weeklong unit, the system atomizes topics so completely that the question of mastery becomes not scalar but binary. Either you know that oxygen is the most abundant element in the Earth's crust or you don't; there is no in-between.

In one sense, technologies of this sort are still a long way from being even a plausible replacement for human teachers. For the moment, the thought of a history class taught by even the best available chatbot—say a circa-2020 Amazon Alexa—is laughable; job-security-conscious teachers would seem to have little to worry about. But then again, machine-learning software is so different from the forking-branch software of yore that the idea of a major breakthrough seems a matter not of *whether* but of *when.*

In the coming years, we'll be seeing plenty more from whence Watson has come. The company Squirrel AI, based in Hangzhou, China, for instance, which helps students prepare for China's year-end standardized tests, already serves as many students as all of New York City's public school system. It slices its school subjects almost absurdly finely—reducing middle-school mathematics, for instance, to 10,000 unique instructional points, which, Pac-Man-like, students gobble up each according to her own personalized path. In the future, "human teachers will be like a pilot" of a plane on autopilot, the company's founder, Derek Li, told MIT's *Technology Review*, operating mainly as a fail-safe while providing significant emotional reassurance to the passengers.

Most examples of this sort of inside-out, machine-learning-powered edtech apply to STEM fields, since their hard edges and unambiguous questions lend themselves more readily to automation. This isn't universally the case, however. In fact, machine learning has made significant inroads, surprisingly enough, in the field of essay instruction and grading. It's been a tale of some promise and no small degree of peril. Machine-learning algorithms, trained on human-graded essays, can theoretically cue in on whatever human graders find good or objectionable about the essays they encounter, grading huge numbers of essays more or less instantly. The GRE, for instance, the major graduate school entrance exam in the United States, utilizes a machine-learning essay-grading system known as "e-rater." An internal 2018 study, commendably made public by the GRE's parent organization, delved into what makes e-rater tick (remember, the workings of such systems remain an unknowable black box, probe-able only from the outside). The results weren't encouraging. For one thing, it appeared to reward longer essays, regardless of how effectively students used their extra verbiage. (In fact, some students have begun copying in memorized passages of "shell text" that, though utterly unrelated to their essay prompt, add length and boost scores.) Far more concerning, the system assigned lower marks to African Americans' essays, in particular, than did human graders. It was possibly (the authors of the study can't say definitely) the result of an issue well known in machine learning: a hint of bias in the data used to train a learning algorithm becoming vastly amplified once the algorithm is operating on its own. In the case of e-rater, human trainers had accidentally created what appeared to be a racist algorithm.

Machine-learning-powered education software, in some instances promising and in many others downright dystopian, makes the strides taken by MOOCs look almost like baby steps. In MIT's development of some of the earliest MOOCs, we recognized that we didn't know everything about learning, and so we insisted on keeping humans in the equation: human lecturers behind the videos; human moderators on the message boards. True, at the scale we were contemplating, such cozy, face-to-face tactics as discussion sections became difficult to pull off. But still, if there were some unknown, vital nutrient necessary for education as we knew it, perhaps with humans in the online mix we could still preserve it. The food we were serving up might be prepackaged, but at least it had once grown in Earth's soil. Personalized, machine-learning-powered education programs, by contrast, are more like a lab-grown, synthetic food, guaranteed to contain every nutrient that's known, which is all perfectly fine—unless, that is, it turns out that it lacks an unanticipated nutrient (or contains an unanticipated toxin).

And anyway, it really isn't all that hard to imagine a few ways a machine-learning-led progression through school could lead to educational deficiency. By breaking down school subjects into impossibly small points, a machine-learning algorithm could easily account for every fact in a curriculum, but not integrate those facts properly during instruction. Students might thus learn calculus, but not learn to *think using calculus*. Worse, if a machine-learning algorithm were trained specifically to optimize standardized test outcomes, the system would risk tripping into cognitively harmful study traps such as massed learning, which leads to good short-term test results but poor long-term retention. And then on top of such mechanistic cognitive issues, there still remain all the now-familiar rejoinders from outside-in educators: even if a student's rate and sequence of learning is customized, for instance, that still doesn't take into account what motivates her, or makes her curious.

I was mulling these and other problems over when Susan Silbey, then the chair of the MIT faculty, told me I needed to talk to one of her PhD advisees.

THE DELTA PROBLEM

By this time I'd taken on a unique role at MIT. Exploring how learning works, and how best to promote it, remains one of the most pressing and fascinating questions facing humankind. It should have come as no surprise, then, that *many* of MIT's departments would find their compass needle of inquiry drawn toward learning at one point or another. The pursuit pulled together scientific disciplines in surprising ways, combining the forces of mathematicians and molecular biologists, linguists and computer scientists, psychologists and physicists. Broadening the discussion to include the art of education only looped in more disciplines, bringing to the table economists and social scientists, teachers and historians and artists.

It was all almost too big to even think about in one sitting, let alone do something about—and yet, we *had* to do something. Wherever any of the above research disciplines intersected with learning, researchers were always there, at the front lines, hacking away at the unknown either from the outside in, the inside out, or somewhere in the middle. We needed some way, if not to physically bring all of these researchers together in one room, then at least to create a framework where their work could coexist and build on itself.

And so, in 2016, MIT announced that it would be creating a constellation of interdisciplinary groups all falling under the auspices of a new unit: Open Learning. Today, Open Learning is broken into two halves: research and development. The research side includes MIT's Integrated Learning Initiative, or MITili (pronounced "mightily"), led by John Gabrieli, the renowned neuroscientist. MITili is our Penn Station of learning and education science, bringing together research threads from all different directions. On the development side, meanwhile, can be found the Office of Digital Learning, which produces online learning media (including MITx courses) for learners on campus and off. And J-WEL: a separate, newer effort funded by the charitable foundation Community Jameel, which gathers together best educational practices from around the world.

The carrier pigeon flitting between all of these fonts of wisdom, improbably enough, is me: the guy who so enjoyed his university courses, he took them twice. The position is essentially sui generis, with no real equivalent at any other major research institute that I know of, and it's a tremendous honor to occupy it. The task of pulling our disparate research threads through the needle's eye of real-world practice is sufficiently daunting that I turn to the help of experts whenever possible. And so, when Susan Silbey, an expert in the dynamics of complex organizations, told me to talk to her student, Marc Aidinoff, I immediately opened my schedule.

Aidinoff had a story to tell. As a PhD student at MIT, he had joined Freedom Summer Collegiate, an organization that brings doctoral candidates to underserved areas of Mississippi to lead seminars for college-bound high school graduates. His doctoral work, a history of how the rise of networked computing has influenced U.S. social policy, seemed at best tangentially related to this mission, but what he observed in Mississippi turned out to be one of the more shocking intersections of computing and government he had ever encountered. He began to make repeat visits, now as an ethnographer, seeking to understand what edtech was doing to schools there.

"It's really hard to get teachers to the Delta, and the schools are particularly poor," he explained—and racially divided. Worse, the region is suffering from an extreme teacher shortage, exacerbated by the inability of schools to offer nationally competitive wages. In 1997, the situation was dire enough that the Mississippi legislature passed a so-called Critical Teacher Shortage Act; today, the shortage in the region is six times worse, and it disproportionately affects poorer districts with a higher proportion of African American students.

To make up for the shortfall, districts have begun turning to technology. Some of these fixes are as simple as they are distressing. For instance, Aidinoff described one high-school-level Spanish classroom that, in the absence of a teacher, simply watched a video feed of a classroom the next school over. Others are more sophisticated. It's become increasingly common to see Delta-region students sitting in classrooms, navigating digital courseware on Chromebooks without an accredited teacher to guide them. Instead, classrooms are staffed with "facilitators" who lack content knowledge. In a typical classroom period, according to an exposé produced by *Mississippi Today* and the independent news nonprofit the Hechinger Report, students access the online portal provided by the edtech company Edgenuity, or one of the state's seven other approved online vendors. The program guides them through sub-lessons broken up into segments labeled "Warm-Up," "Instruction"—a lecture video—"Summary," "Assignment," and "Quiz"; larger tests come at the end of units.

To be fair, these schools are not using this product as recommended; it is "designed to work hand in hand with teachers," an Edgenuity company representative said in an email. This approach makes a good deal of sense: online learning tools are almost always best used in conjunction with human instruction. But at the same time, the very existence of online-only options can, in cash-strapped school districts, create ideal conditions for misuse. Without a teacher in the room, for instance, student questions can languish unanswered. "We're just going over some of the facts that we should know and it's kind of difficult," one student told *Mississippi Today*. "If I've got question to ask, how you do that? I can't ask the computer and the teacher that up in there, she don't know it."

Just as concerning, without a content-savvy figure in the loop, the practice of teaching directly to standardized tests, already all too common, can become clad in software iron. One reason a school system might turn its limited budget over to digital courseware companies is that they sometimes promise good results on standardized tests. "The principal is faced with a choice between a teacher, or a company that's selling either software or packets, often with a guarantee of some kind of test rate," said Aidinoff. As a result, "it's very common that, for the entirety of high school, you never hold a book in an English class." Rather, the digital courseware feeds you passages that resemble the passages you'll encounter on standardized tests. "You're basically doing SAT prep in some way, and often repeating the passages multiple days so that you can get better," he said. "The students really repeat lessons very, very often."

Perhaps the most concerning side effect Aidinoff observed was how the program transformed the remaining adults in the room—the non-teaching "facilitator" or "school resource officer"—into stand-ins for authority, not knowledge. "The expertise the 'teacher' has is disciplinary and nothing else, really," he said. "That, to me, is the question that most of these programs miss. They forget that there's going to be a person. And I'd generally prefer the person wasn't functioning as a cop or, literally, a 'school resource officer.' "

—

I still believe in edtech. I believe in its potential to improve human flourishing: to loop more of us into the world of learning and carry us farther along profound educational journeys. To amplify the power of teachers. To nimbly apply both inside-out and outside-in research findings. To make learning user-friendly across the board, and therefore more attainable for a wider variety of brains hailing from a wider variety of backgrounds.

But it's now clear that any such deliverance will be complicated. As studies into the effectiveness of MOOCs continue to show, despite recent, staggering increases in worldwide mobile internet availability, access alone can't guarantee that anyone will find her way to online education. Roughly half of all people on Earth now use the internet; in India alone, the internet-using population has multiplied by a factor of *thirteen* since 2007. And yet, there is a whole host of reasons why someone with a smartphone or connected laptop might not take an online course, ranging from limited free time and energy to the local absence of jobs where a head full of knowledge can be put to use.

One especially important factor is the question of whether there's a human being with content knowledge in the mix. The students in the Delta region are only part of a larger trend. A sizable 2019 review study conducted by MIT's J-PAL—a sister organization to J-WEL that is focused on global poverty, co-founded by recent Nobel laureates Esther Duflo and Abhijit Banerjee—determined that students in traditional classrooms tended to outperform those in online-only courses, although the online-only option was still certainly better than nothing. However, students in online courses who *also had an in-person instructor* performed just as well as those in traditional classrooms, and when those online systems created personalized student pathways (particularly in math, often aided by machine-learning mechanisms), some students performed as well as those receiving private tutoring—the gold standard for personalized education.

Even Battushig Myanganbayar, the MOOC poster student, had in-person help: not just his MIT-alum principal who first directed him toward Course 6.002x, but also a visiting teaching assistant from Stanford, who was a friend of the principal. As Justin Reich, the ed researcher who has led most of the recent independent MOOC studies I've mentioned, has argued, Battushig's story "isn't a story about an individual, it's a story of the MIT alumni network and a community of people working together to raise the quality of education in Mongolia." The headline of the *New York Times* article about Battushig called him "The Boy Genius of Ulan Bator," Reich pointed out, but "the story could have been called 'The Boy Genius of Ulan Bator and His Very Well-Educated Mentors.'" Although any MOOC is far, far better than no instruction, its powers magnify geometrically—especially, I think, for younger students—when you introduce a human being to whom students can pose questions as they arise.

The J-PAL study does not draw distinctions between the different types of artificial "cognitive agents," as the study calls them, undergirding digital courseware. As the more effective, machine-learning-endowed exemplars continue to distinguish themselves, we may see students, particularly older ones, striking forth as tech-augmented solo learners, and conquering at least some topics as effectively as those with personal tutors. But no amount of technological advancement will outweigh the most significant factor that can make or break tech-enabled learning: its broader social and institutional context.

Take Course 2.007. One of the harshest lessons it teaches its roboticists-in-training is that even when a technology always functions perfectly in the lab, if it doesn't work in the wild, then it's not a successful technology. In the case of edtech, "the wild" must be understood to include not just well-heeled schools, but also cash-strapped institutions that find themselves forced to use software—even against the express advice of its creators—to replace teachers.

Perhaps what disturbs me most about the Mississippi example is not just how resoundingly the technologies in question fail students in the Delta region. It's the fact that *the very same technologies*, wielded as a tool by skillful teachers in relatively munificent districts, could easily have a reverse, beneficial effect. That may sound like it's not much of a problem, but it is: Such a technology could act like a sieve, benefiting the haves and harming the have-nots. Assuming your underlying goal as a technologist is, like mine, to raise everyone up, *especially have-nots*, then such a technology is worse than a dud. Instead of recalibrating the educational winnower to be more forgiving, such a technology has the potential to reinforce its worst biases.

POISONING THE WELL

Educational technologies that harm students today have the potential to poison the well for better technologies to come. Two hundred years ago, in the Western world's first encounter with personalized, mass education, the existence of substandard schools run by Andrew Bell's acolytes and imitators—especially Joseph Lancaster—lent strength to the odor of charity and poverty already clinging to the monitorial approach. Such schools, it became clear, were acceptable only for *other people's kids*.

It probably didn't help that Lancaster, in particular, developed a reputation for putting most of his energy into creative punishments for bad behavior. The role of discipline and punishment in Lancaster's failed system takes on new, queasy-making life in the context of today's Delta region schools, where students work under the watchful eye of classroom disciplinarians, not teachers. It's a recipe for resentment toward all things related to school.

It's difficult to get a sense of precisely how many schools across the United States, let alone the world, are replacing their teachers with digital courseware; there are too many independent school districts and too many companies servicing them. I've asked a number of onlookers in both the edtech industry and associated think tanks about the issue, and they have estimated that the pattern that has appeared in the Delta region probably remains fairly rare. However, programs that are mechanistically similar, even if they fit differently into the larger education landscape, are popping up with increasing frequency. "Credit recovery," for instance—an umbrella term for strategies that allow high school students to make up classes they've failed or missed—often hinges on online courses. Again, it's hard to say exactly how prevalent the practice is, and the specific approaches used in online credit-recovery courses vary wildly, as, presumably, does quality. But it does appear to be widespread: in 2011, 90 percent of U.S. school districts engaged in some kind of credit recovery.

As online learning continues to make inroads into school districts, classrooms, and homes, I'm growing concerned that it will trigger a public outcry before its full potential can be realized. One volatile point that's only begun to come into focus is the matter of how student data will be used. I don't mean just the normal privacy concerns we all have, with unaccountable entities tracking our every step and selling that information—although the thought of a shadowy marketplace for children's behavioral data is especially creepy. As concerning as that is, I'm even more worried about how online student data can be fed into the educational winnower—and how that will in turn affect the well-being of students trying to learn.

Traditionally, a student's academic progress has been reflected by a relatively sparse data set, collected mainly on test days and at the conclusion of big assignments. But the flip side of the promise of hyperpersonalized education is the specter of highly granular, ongoing assessment: the ceaseless weighing, sorting, and ranking of students. In theory, a course of online study could conceivably rank students according to everything from their study habits to their mouse cursor movements to (far creepier still) their facial expressions. (MITx does none of these things, by the way.)

On one hand, a minimally creepy, continuous assessment regime might represent an improvement on how we currently sort students. The twenty-eight-day admissions piscine at 42 operates somewhat along these lines, and it creates a far more detailed picture of a student than do test scores and letters of recommendation alone. But the idea of making every moment in a student's day or week a potentially sort-worthy event is deeply troubling. I've already suggested that a certain type of student flourishes in our traditional educational winnower; perhaps a shift from sparse to continuous surveillance wouldn't solve this problem, just select for a new personality type. More concerning, if a student's every click, every keystroke, every failed attempt on a problem set redounds to her permanent record, then she'll never have true freedom to take risks and fail, like the students in 2.007 do. If any future purveyor of a hyper-personalized education system wants to avoid sacrificing learning for the sake of sorting, they must, at the barest of minimums, permit students to make mistakes that are not recorded, never carved into digital stone.

—

If machine-learning-enabled personalization threatens to merely replace one reductive winnower with another, you might reasonably wonder whether it's a technological avenue worth exploring at all, or if we should just leave it well enough alone. As tempting as that option might seem, however, it would be a major mistake, if only because machine learning's theoretical instructional capabilities range far beyond the horizons of today's classrooms.

To pick one fascinating example, take the work of Fox Harrell, a true polymath who holds a joint appointment at both CSAIL and MIT's Comparative Media Studies Department. Harrell produces interactive, multimedia works of computer art, often powered by home-grown machine-learning systems, not to mention Harrell's deep understanding of cognitive linguistics: one of the more fractious battlefronts of the cognitive revolution. An important assumption underlying his work is the idea, borrowed from the cognitive linguist George Lakoff, that we all rely on nested structures of cognitive metaphors to make sense of the world. Some of these are simple and apparently universal, with deep sensorimotor roots (for instance, the idea of "more" represented by the direction "up"), while others vary by culture, and still others are personally idiosyncratic. According to Lakoff's theory, these metaphors are a major part of what makes each of our experiences of the world unique. Harrell's software artworks are designed to build bridges across personal and cultural metaphors, to help you begin to see the world through the lens of someone else's constructed reality. It's a heroic project that is still in many respects getting under way, and yet it's an important proof of principle. Technological systems need not only oversimplify human experience; they can also *complicate* one's education in a way that might not otherwise be possible. With the promise of such powerful techniques to look forward to, a general turn away from machine learning in education would be unfortunate, to say the least.

That's not really likely, in any case. What's more probable, to paraphrase the science fiction novelist William Gibson, is that as the machine-learning-enabled future arrives, it will continue to be unevenly distributed. Necessarily, the more a technology diverges from how education is traditionally done, the less compatible it will be with existing institutions. It was for this reason, in fact, that IBM Watson diverted the bulk of its energy from K–12 into higher education, where the rules are less rigid, and professional education, where there are no rules at all. ("It is the Wild West," said IBM's Alex Kaplan.) Suppose cutting-edge edtech continues to develop in this way: forced to grow in the cracks between existing institutions, in the mold of Sep Kamvar's Wildflower Montessori schools, or else in the wide-open spaces of developing countries. Such a world might well see what's known as a technological leapfrog effect: where developing regions race ahead of their more developed (but also more entrenched) brethren. Leapfrogging countries skip the landline stage of telephonics and adopt mobile internet, for instance (something I've personally observed in my own family in India), or jump ahead of clunky electronic money transfers and set up seamless mobile banking. Could less-developed parts of the world, and nontraditional students everywhere, armed with new modes of technology-assisted education, leap ahead of their more entrenched counterparts?

"It's already happening," said Agarwal, now the CEO of edX. In a Circuits Analysis course at San Jose State University, the classroom pass rate was 55 percent in 2012. "When they used the edX class on campus, the pass rate shot up to 91 percent. It leapfrogged." He continued, sotto voce: "But we just try to downplay, because it's very threatening." (Meanwhile, adding gasoline to the fire, a concurrent experiment with a rival MOOC provider did not go nearly as well, which caused San Jose State's faculty to revolt—justifiably, in my opinion.)

Technologists are now hard at work, designing systems threatening—no, that's not the right word—*promising*, sooner or later, to spread quality education more widely than ever before, while improving on the status quo wherever possible.

But in the past several years, even as technologies have arrived with the potential to challenge the caprices of the educational winnower, the winnower has proven resilient: bending those technologies that can be bent, and outmaneuvering those that can't.

And so we find ourselves in an uncomfortable resting place. Whether your preferred approach is outside-in, inside-out, or some hybrid of the two, improving learning alone is not enough to fix education. There still remains a system at work, hell-bent on designating students wheat or chaff based on factors that still have little to do with their potential as learners, let alone as human beings.

Perhaps, if we're ever going to truly help people everywhere tap their unrealized wealth of potential, it is time to address the winnower head-on.

- IX -

THE SHOWDOWN

It was the second and final day of the Course 2.007 robotics competition, and Woodie Flowers, the guest of honor, took to the podium to kick off the event. An oversized cutout of the Death Star loomed behind him as he surveyed the crowd.

Flowers was normally a sunny personality, but today he had not come to deliver good news. The state of knowledge in the United States, he soon made clear, had taken a perilous turn.

"In this country, for example, less than half of the adults accept evolution," he said. "We're doing crazy things. We're shutting off the truth that's associated with science. The EPA is purging scientists from their governing structure.

"One major thing that's going on that requires a lot of attention is the wealth divide," he continued. "If your parents are in the top one percent"—he gestured at a slide referencing work by the Harvard economist Raj Chetty—"you have a good chance of being in the top 30 percent. If your parents are in the bottom one percent, strong indication you will be in the bottom 30 percent." He looked out over the crowd. "That's not the way it's supposed to work."

Among his audience in MIT's Johnson Ice Rink—a hundred-odd Course 2.007 students and several times that in supporters and alums—sat the thirty-two competitors who had fought their way to the final event. "You guys have to save us from ourselves," Flowers said.

It was late May: that sleepy, valedictory moment in the academic calendar when speakers attempt to slip the starched overcoat of generational responsibility onto younger, unsuspecting shoulders. But if anyone had the right to demand action from his course's students, Flowers did, because in a small yet quite literal way, he'd equipped them for it. In Course 2.007, he had created an educational oasis where students could rapidly contextualize and apply engineering knowledge; from which they emerged ready to exert their will on their surroundings through the medium of robotics. As Amy Fang, one of the contenders still standing on this final day of competition, later said: "It's so tangible, the things that I've learned, and that I can actually apply in the world."

The audience, meanwhile, was anything but sleepy. In fact, it was singing with anticipation. Of the thirty-two competitors remaining, sixteen had scrabbled their way to today's final competition via yesterday's ruthless round robin, into which a total of ninety-five competitors had entered. These survivors had spent the night strategizing and making last-minute repairs. Meanwhile, the other group of sixteen, who for the most part owned the top robots in the competition, had yet to be tested in public. They had earned their place via the "Ladder": tournament slots the professors had set aside in the weeks prior for those best able to score points in the lab. This latter Ladder group included Alex Hattori, the former BattleBots competitor; Z, who had unveiled a robot capable not only of spinning both cylindrical thrusters but also of delivering a smaller robot to the X-wing's upper deck for the purpose of sabotaging his opponents; and—surprisingly enough, given his early struggles—Brandon McKenzie, the swimmer.

The clear favorite, though, was Tom Frejowski, a lanky, reserved Chicagoan with a long history of tinkering in his bedroom workshop. Like many of this year's top contenders, he'd designed a fork-like device capable of reaching into the foursquare buttonholes of the X-wing's lower thruster and spinning it. Tom's fork, however, was uniquely reliable: It was attached to a pliable rubber coupling that allowed the thruster to spin even if the robot were imperfectly aligned with the thruster's face. This neat bit of design, combined with some highly strategic coding, gave Tom an enormous edge. Each full, single-elimination competition round would take a total of two minutes, but the first thirty seconds were designated solely for robots capable of autonomous action, with no human at the controls, and any points scored during this period would count double. Then a tone would sound and ninety seconds of radio-controlled play would begin. At this juncture, most students would be just setting out on their quests for points, but not Tom: By the start of the radio-controlled period, he would have already banked 416 points for spinning the lower thruster autonomously. Then—and this was the real trick—Tom's robot would *continue* spinning the thruster as the autonomous period gave way to radio-controlled play, instantly earning him an additional 208 points. As a result, by the thirty-first second, Tom would already be sitting pretty with a staggering 624 points on the board—a feat made all the more impressive by the fact that his robot was, in the words of professor Amos Winter, "dead reliable." A handful of other robots could theoretically amass comparable point totals, but none so consistently.

There was one contingency Tom hadn't prepared for, however: a human one. His name was Richard Moyer, and he had been forced off the Ladder at the last minute. As a result, he'd had to fight his way into today's competition via yesterday's round robin, which made him something of a dark horse entrant. But in his very first round yesterday, jaws had dropped to chests and eyes had fixated on the scoreboard. His robot, like Tom's, sauntered up to the thruster during the autonomous period. Then it fired up a motor that sounded like a lawnmower—louder than any other 2.007 robot by a wide margin. The rumor mill soon made clear that Richard had customized one of the competition's provided motors beyond all recognition; it was the only one of its kind.

By the end of his round, Richard's score was glowing in red: 912 points. Tom wore a tight smile. Still, Brandon, seated nearby, had wondered aloud whether Richard had sealed his own fate by designing a robot that relied on a custom-built motor. "That could burn out," he'd said, a hint of optimism in his voice. "He has no way of replacing it."

—

Even as the final thirty-two competitors took in Flowers's words, the mood was one of eagerness, not fear. Regardless of what happened today, nothing *bad* would happen to anyone. No one's future would suffer if her robot fell apart in a pile, because the semester's-end contest had no direct bearing on students' grades. In effect, the competitors had been operating in a safe zone of sorts: free to take risks, make mistakes, and learn from those mistakes. This sanctum, where excitement outweighs fear, is actually what enables Course 2.007's other pedagogical achievements, including how it motivates curiosity and contextualizes engineering fundamentals through real-world practice. Rather than add to the winnower's existential threat, Course 2.007's competitive, hothouse atmosphere somehow makes its lethal apparatus feel like an afterthought.

In Brandon's case, that equated to a chance to scrap his erstwhile scissor robot and start over. "I decided to revisit the drawing board," he said. His new robots—plural—looked somewhat rickety, but he had adopted Z's direct-drive approach for his new, thruster-spinning fork, and there was no denying that they made the thruster hum.

Safe in 2.007's embrace, its students are given an experience that is, I think, as true to the founding ideals of MIT, set down way back in 1861, as they get anywhere on campus. The principles of *Mens et Manus*, "Mind and Hand," predated E. L. Thorndike's influence by thirty-five years, as well as such institutional fingerprints as credit hours and GPAs, even federal accreditation of universities. Most importantly, *Mens et Manus* predated our supposedly meritocratic, entirely ruthless, cradle-to-grave system, which was built to enable winnowing as much as teaching.

Today, as I ponder how to avail a truly effective education to as many people as possible, I keep returning to Course 2.007. The way it separates school's winnowing function from its teaching function—stripping apart two aspects of education that have been glued together for over a hundred years—hints at at least one viable way to rethink the winnower and make education substantially more inclusive.

THE MICROMASTERS

Taking the broadest possible view, recovering vast quantities of latent learning potential—not to mention improving a whole lot of lives—is as simple as optimizing instruction while also improving access to it. As we've seen, however, these two levers often work at cross-purposes. Some of the most successful attempts to inject modern cognitive science and personalization into educational practice come with a lamentably limited reach: extending only to small, experimental schools, say; or student demographics outside traditional educational institutions, such as midcareer professionals or kids growing up in rural Mongolia. Such populations may amount to a new, untamed "Wild West," as IBM's Alex Kaplan put it—an exciting proving ground, to be sure, but one that leaves behind everyone already inside the schoolhouse walls.

The challenge before us, then, becomes something of a two-handed piano improvisation. On one hand, how do you best ensure that the newfangled educational opportunities reaching those "Wild West" populations are actually useful; that those students can plug their new knowledge back into society in ways both personally remunerative and beneficial to their communities? And on the other hand, is it possible to expand the borders of this Wild West to include more learners every year?

I can't tell you exactly which strategy, if any, will win out for every age group, in every discipline, everywhere in the world. But at the very least, I can tell you what we're doing about it at MIT.

Ultimately, higher education is where MIT wields the most influence*—and as it happens, certain types of higher education are particularly ripe for re-wilding. By 2015, MITx and edX were already off to a strong start when I began to get the indescribable feeling that we had a chance to zag while everyone else was zigging. At the time, MOOC providers, edX included, had begun to dabble in awarding badges and completion credentials to students, which in theory allowed them to prove to employers that they'd learned something. In practice, however, issuing a new, made-up credential came with the same problems as printing a new currency: They could only achieve value in the wider marketplace by proving that they were valuable, a catch-22.

We had one coin of exchange that others didn't, however. We

* Many millions of children and teens routinely interact with MIT more indirectly, however, through such intermediaries as Scratch, Khan Academy, Quizlet, MIT App Inventor, littlebits, and Guitar Hero: all invented either in-house, by alumni, or even by highly successful dropouts!

could back a new online credential using the bullion of a traditional, on-campus MIT degree—perhaps the Institute's most valuable asset.

We wouldn't mess with MIT's bachelor's degree—in many ways a sacred object—nor with our PhD. The master's degree, however, was intriguing, featuring a number of arbitrarily stuck-together elements that, with a little care, could be teased apart. What if, I wondered, we were to crack a master's program in half, putting the parts most suited for online learning on the internet and reserving for campus those parts requiring face-to-face interaction? Those steps alone wouldn't actually be all that radical—approaching a flipped classroom, essentially—but the second part of the scheme would challenge the established order of things.

An online first half of a master's degree, I believed—essentially a highly intensive semester's worth of courses, delivered online over a year or more—was a course of study worth certifying in its own right. This credential, which we eventually named the MicroMasters, could also do something else, however. It could serve as the primary *admissions* criterion for a second, on-campus half of a full master's degree. In such an arrangement, we could educate students by the hundreds of thousands online, and then, for those willing to go further and able to distinguish themselves, offer a single intensive semester on campus. With their online and on-campus coursework combined, they would earn a full master's degree.

Bully for them, but the hidden beneficiaries of this setup would actually be the online-only MicroMasters holders who never set foot on campus. Precisely because MIT would be treating that initial salvo of online courses as credit-worthy for the couple dozen students jumping from the internet to an on-campus master's program, it testified to the value of tens, perhaps hundreds, of thousands of stand-alone MicroMasters' degrees out in the larger world. Our online credential would be backed by on-campus gold.

Meanwhile, depending on the specific master's program we picked to pilot this new progression, using the MicroMasters as an on-ramp could substantially broaden the applicant pool for the full master's degree. In theory—and this would be the real test of the entire scheme—infusing a traditional, on-campus master's program with the fresh blood of MicroMasters students from different walks of life, all around the world, might, if anything, *improve* the quality of master's graduates we ultimately released into the wild.

The biggest question was who might be willing to crack their department's master's program open—but this, at least, was an easy call. I went straight to Yossi Sheffi, who heads MIT's Center for Transportation and Logistics. Sheffi's unit offered a master's degree in supply chain management, and I knew him well from my RFID work, which involved tinkering with industrial supply chains. Sheffi's team already couldn't churn out supply chain professionals fast enough to meet industry demand, which meant, we reasoned, that there was a wide world of serious companies that would be all too happy to recognize a scrappy, new supply chain credential, assuming the quality of our graduates held up. Sheffi himself, meanwhile, was the perfect co-conspirator. A serial entrepreneur who had brought five successful companies into the world, he was ready to move fast, and as the godfather of a relatively new master's degree, he knew the sorts of challenges awaiting us as we floated the MicroMasters scheme to the larger MIT community.

When I broached the idea, Sheffi became immediately enthusiastic. "It's the right idea at the right time," he later recalled thinking. He roped in Chris Caplice, the Center's executive director, who would design the new credential and teach several of the courses involved. Sheffi told him, "Chris, we are not stopping. We are doing it as if it's a startup. The speed is everything." There would be internal pushback, and understandably so: We were talking about splitting educational atoms—breaking apart what had been the Institute's most fundamental unit of education. But Sheffi declared that "we're going to do it and then I'll get MIT to come around."

And that's essentially what happened. The process wasn't exactly easy or straightforward, but both the administration and the faculty gave the idea a fair hearing, and eventually things began to flow more smoothly. Sheffi, Caplice, and Eva Ponce, a senior logistics researcher who ended up running the day-to-day operations of the MicroMasters, took the existing slate of three supply chain MITx courses and added two more, creating a five-course sequence. And students who had been taking those initial online supply chain courses, going back all the way to 2014, found themselves presented with a shot at a MicroMasters, and possibly even a full master's on campus.

Paulina Gisbrecht, a native Muscovite who had spent most of her early career in Germany working for GE's thermal services unit, was one of them. She had minored in logistics at university and found herself introducing some newfangled logistics principles to her unit at work, which sometimes ran into problems sourcing spare parts for massive steam turbines. Shortly after completing a business school certificate, she began looking for a follow-up course of some kind. A friend pointed her to MITx, where she discovered a brand-new slate of supply chain MicroMasters courses. She began to read about the program and realized the courses could teach her "a lot of analysis stuff and really quantitative things"—the nitty-gritty of supply chain dynamics that her relatively cursory undergraduate coursework hadn't dealt with. Unlike some other online courses, however, the new MicroMasters courses were only offered at set times in the academic year. She checked when the next introductory course began. It was slated to start on that very day, in just three hours. "It was like, 'Thank you, God. You gave me a course today that I was looking for,'" she said. "That was a sign."

Srideepti Kidambi, Gisbrecht's eventual classmate, found her way to the same place via a different path—a surprisingly familiar one to me, in fact. Hailing originally from India, she earned a bachelor's degree in mechanical engineering from BITS Pilani (a prestigious private technical university), and, after graduating, she found herself working as a supply chain engineer for Schlumberger, the selfsame company where I'd found my first job after university. She even underwent a training period on an oil rig. Afterward, she bounced around the world for a few years: Pune, in India; to Houston; to Singapore; where she found her way into the consulting world. Then she and her husband started a family. "The consulting life with two kids didn't really go very well together," she said, and she left her job. Around the same time, her husband accepted one in New York City, and the family moved yet again. Soon enough, she was ready to return to fast-paced, high-impact work, but found the prospect of her former travel schedule untenable. She mulled going back to school, and even looked into a traditional master's in business administration, but the timing, and the time commitment, never felt right.

The MicroMasters, however, "was in that sweet spot for me, where I could actually work and study at the same time, and didn't have to travel."

While starting her first online course, she began interviewing for jobs. "I was looking for companies that were more in the startup phase," she said, where "you could actually drive a lot of change. So that's when I started looking at Rent the Runway, Blue Apron, Shapeways, a couple different companies in different industries."

The MicroMasters proved "a very good conversation starter" in the interview process, she said. The chief logistics officer at Rent the Runway, in particular, was blown away by her progress; it turned out that he had been attempting the course in his own spare time and knew how demanding it was. "He didn't end up finishing it," she said, "but he was like, oh my god, I can't believe you're actually going through all those sums, you're actually doing all the math." He ended up hiring her.

The supply chain MicroMasters was, indeed, extremely rigorous—and demanded a lengthy commitment for anyone planning to go the distance. Each of the five courses took about three months. From the start, Kidambi had hoped to fight her way into the blended on-campus program, but "my family in general was not sure if I would actually be able to finish the whole thing," she said. "I was like, you know what, let me give it a shot. Maybe if I like it, I'll continue it." Instead of adding yet another chore to her already busy days, however, the courses turned out to be "intellectually very stimulating for me and very fulfilling."

Gisbrecht had a similar experience—in the beginning, anyway. The MicroMasters "is a *lot*," she said. Prior to signing up, she had already booked a vacation to the Philippines, and so her third and fourth week of the first course found her studying on vacation. "Everybody was diving and I was sitting in the room," hard at work, she said, "but it was worth it. I just didn't want to miss it. I did the first fundamentals course and I did well."

In fact, she aced the first two courses. "Maybe, maybe I will have a chance," she thought, of making the cut and joining the blended cohort on campus. She wrote in her diary that she planned to get at least 90 percent in her courses. At the time, "I didn't know that I could do better than 90 percent," she said, chuckling.

An ocean away, as Kidambi moved through the same opening courses, her boss at Rent the Runway gave her the time she needed to keep up. "He was like, 'Okay, I know you've got a midterm coming up. It's okay, take some time off, go do it.' Because he saw how intense it was. That really helped me," she said.

Around the same time she approached the second or third course, "Rent the Runway started thinking about growth and expansion," she said. The company, perhaps unsurprisingly given its name, rents out designer clothing, and its business model hinges on its ability to send out, receive, and clean its garments—which calls for an extremely efficient turnaround. Every one of the fast-growing company's garments, however, passed through a single facility in New Jersey—"the world's largest dry-cleaning facility," Kidambi said—a classic recipe for a logistics bottleneck. A second facility was clearly overdue. "From the concepts I learned at SC2x"—a highly analytical supply chain design course—"I actually built a network strategy model," she said. She showed it to her boss, who "took it to a larger audience, the COO, the CFO, and everyone." In the end, the company adopted her plan to open a new distribution center in Arlington, Texas, which would handle 40 percent of the overall flow of garments.

Kidambi was doing so well at her job that when she expressed her continued desire to attend the on-campus master's program, her colleagues expressed confusion. "People were like, 'You don't even need a master's. You're good the way you're performing,'" she said. But an alumnus of MIT's traditional master's program pulled her aside. "He was like, 'Forget what everyone else says. Just go do it. I think you're going to regret it otherwise.'"

What neither Kidambi nor Gisbrecht knew, however, was their chances of admission. "I had no clue," said Kidambi. "I didn't know whom I was competing against." Gisbrecht, meanwhile, hit a snag in the third course in the series. The final exam had "one very tricky question," she said, "so, in one course I actually scored not that well. A 'B.' I thought, that's it. I have no chance." She wasn't the only one stymied by the question, she said. "I know a couple of people who were so active before," on the course's message boards, "and who just jumped off because they were so destroyed by this."

That same week, however, she received an email from MIT asking her to sign up to be a community teaching assistant: someone who could help other students who were struggling. She took it as a signal that she still had a chance, and gladly assented.

All told, the entire MicroMasters component of their master's degrees took both students sixteen months, including time spent studying for a single, high-stakes final exam, proctored via webcam. "One guy," said Gisbrecht, "was working in Sudan at that point of time, with literally no connection to internet." So, rumor had it, he claimed his company's entire bandwidth. "He just plugged in and took all the internet. He was just like okay, tomorrow I'll have to deal with it, but now, I have to do my exam."

"On the first of August, we received our admission," she said.

"I was like, I'm going. I'm very excited," said Kidambi. "For a couple of years, I couldn't get myself to go to school," but now, "I was like, I just want to go and get it done," she said. "All my hard work paid off."

A BETTER WINNOWER

Undeniably, even with the MicroMasters in place, a winnowing is still occurring in the supply chain master's program. At the moment of writing, 300,000 have registered for at least one supply chain MicroMasters course, 30,000 of whom have earned a lower-level certificate for completing said stand-alone course, and 1,800 of whom have earned the larger MicroMasters credential. Of that 1,800, almost 15 percent have applied for the 40-odd spots reserved in every year's on-campus master's cohort for online learners, known as blended students, who join midway through the program's two semesters. The applicants who come in from the online cohort, however, are such high achievers, said Eva Ponce, one of the program's principal administrators, that it's likely that students are actually self-winnowing, rather than wasting time on an unlikely application.

But the bestowal of a high-value MicroMasters on even students who don't find their way onto campus means that the year-and-a-half-long ordeal is anything but a waste of students' time; so long as they pass their courses, they get a valuable credential for a fraction of the cost of a master's degree: $1,000 total, $1,200 if they choose to sit for the final exam.

And meanwhile, if this new winnowing process is not perfectly meritocratic—it still relies on high-stakes tests, for instance, with their intrinsic drawbacks—it at least comes closer to that ideal than traditional admissions processes, and leaves less human potential on the table.

Part of that story has to do with the very deliberate way Chris Caplice has separated testing-for-promoting-learning from testing-for-admissions. Initially, when MITx's supply chain courses did not lead to on-campus admissions, Caplice treated assessment purely as a teaching tool. "There was rampant cheating, to be honest," he said, but it didn't much matter. The low-stakes testing, designed to encourage collaboration, provide instant feedback, and promote the effortful retrieval of the sort advocated by the Bjorks, was an important part of what made the courses sound from the standpoint of students' cognitive processes. If students cheated, they only cheated themselves.

When those same courses became part of the MicroMasters, however, and therefore eligible for MIT credit, "everything changed. And so that's when the assessment became really critical," he said. An easy mistake would have been to clamp down on the existing testing regime, making it harder to cheat but diminishing the tests' instructional value. Instead, Caplice simply added more tests. Today, the easy-to-cheat, low-stakes tests remain, but so do an ironclad set of midterms and finals, which, thanks in great part to randomization techniques developed by the physicist Ike Chuang, are next to impossible to cheat.

The upshot is that the intrusion of a winnowing function in the course hasn't much affected how the online course's content works from a cognitive perspective. All told, thanks to such touches as the continued presence of learning-centered tests, pause-able and rewindable video, and the fact that the videos are limited to ten minutes in length, the online instruction remains arguably "a better method of teaching certain things than face-to-face," Caplice said.

Tempering the coldness of the ironclad assessments, meanwhile, is the fact that the people running admissions for the on-campus master's degree are the same as those running the MicroMasters. As a result, it becomes possible, when assessing students, to put a single bad test outcome or two into a larger perspective. As Gisbrecht discovered, although a single wrong exam answer dealt a hard blow to her overall course grade, the admissions team saw fit to overlook it. "That's the good point about this online learning," she said. "They actually review all of your performance. If you have maybe one time a bad result, then they say 'Okay how are you performing in the entire course? Oh, you'll perform very well.' "

By the time students finish the MicroMasters, "we know them better than their mother knows them," said Yossi Sheffi. "We know everything: how they think, what they do. So we could choose really the best."

At the same time, the MicroMasters program casts its net over a far wider applicant pool than the traditional admissions process, bringing in a greater variety of so-called traditional and blended applicants alike. Sheffi leaned in and raised his eyebrows. In the master's program, he said, "we have one blended student—we did it without telling anybody—one who doesn't have a bachelor's degree. He was working and a very smart guy. He did great on the online. We said, You know what? Let's take it to MIT, let's see what happens." He laughed. "He was an A student. Aced it." Now, he said, the program publicly states that it will accept highly qualified master's students even if they don't hold a bachelor's degree.

The program's proof, ultimately, is in the blended students' performance on campus. In yet another independent study, Justin Reich and his coauthor compared their performance against traditionally admitted master's students in the supply chain program. Not only did the blended students outperform the traditional students in supply chain courses, but, in courses they took *outside* the supply chain program, they also outperformed the larger population of MIT students. The findings are a testament to the sheer human potential sitting latent, just outside the traditional sight lines of the academy.

—

Today, the supply chain MicroMasters won't only get you into MIT. Although in my biased opinion our master's program remains the top prize, twenty-one other universities on five continents will also give you credit for the online coursework. The supply chain program, meanwhile, was also just the first of many. As of mid-2019, there are fifty-two MicroMasters programs on offer at universities all around the world, four of which are provided by MIT.

Perhaps most importantly, the MicroMasters saga has allowed us to take a highly intentional look at what exactly makes a higher-ed degree valuable—and what aspects of traditional higher education run at right angles to students' goals, or even against them.

There is a truism whispered among administrators at top colleges and universities concerning why students keep signing up for elite schools. The value, the thinking goes, comes in three parts: the stuff you learn, the people you meet, and the fact that you got in. Add in a few more elements of campus life—athletics, say, and parties, and ivy-covered buildings—and you've got a stereotypical sketch of what many American universities offer undergraduates and master's students.

The MicroMasters, meanwhile, prioritizes the "stuff you learn" above all else. It also preserves the "who you meet" as an important, if secondary, concern. (Indeed, at the first-ever MicroMasters Completion Celebration, suspended across a gigantic videoconference in May 2017, small groups of students in cities all around the world called in from house parties, filled with people they'd met online.) It obviously strips away the physical infrastructure of the university, and some of how university life is organized temporally. But what the MicroMasters truly explodes—deliberately, intentionally, pragmatically—is the "fact you got in." In fact, it reveals that "the fact that you got in" was never a unitary element of higher education. It was always two things framed in opposition to one another: a testament to the promise of the admitted student, and the implication that everyone else denied admission is less promising. The MicroMasters dispenses with this Manichaean mode of thinking. In fact, it's possible to be promising in an *absolute* sense, relative only to the rigors of the coursework standing before you, irrespective of the promise of your peers. The core project of the MicroMasters, then, is a repudiation of the perspective we inherited from the first intelligence tests, which normalized one's score against those of one's contemporaries. In the MicroMasters, if you prove equal to the material, you get a certification—no matter how many of "you" there are.

Exclusivity, meanwhile, does persist in the MicroMasters-fed, on-campus master's programs—but no longer as a feature of higher education so much as an unfortunate bug: a side effect of the fact that any face-to-face education program must necessarily be limited in scale. The second half of the MIT supply chain master's sequence, for instance, features a student thesis and a lot of collaboration on the sorts of problems that don't have definite answers, the sort of work that's especially hard to translate to a digital-only medium. As Erdin Beshimov, who oversaw the initial launch of the larger MicroMasters credential, likes to say, we've taken MIT's *Mens et Manus* credo and put the *Mens* online, while keeping the *Manus* on campus.

There are drawbacks to this arrangement: to delivering a corpus's hard facts first, and pursuing complex problem solving second. It's harder to use complexity to trigger curiosity, to motivate learners, and to contextualize knowledge; as outside-in-leaning educators like Mitch Resnick and schools like Ad Astra would prefer. However, at least for adult students doing master's-level work, I think the benefits outweigh the costs. By putting a given field of study's hard knowledge online first, we can teach it to a sizable chunk of the world who might otherwise never get the opportunity.

But why bother, you might reasonably wonder, to keep *Manus* in the equation at all? If *Manus* is difficult to scale online and expensive to run in person, and given that researchers are not even in agreement about whether problem solving *can be taught* as a stand-alone virtue, perhaps we should just cut our losses and stick to teaching hard facts and skills over a broadband connection.

My answer is a simple one: If the ultimate point of education is to create learners who can change their world, then *Manus* is non-negotiable—and access to it must be made more democratic. Whether it takes the form of solving problems on an oil platform (or, more preferably these days, a wind farm), or perhaps piecing together the argument for an essay about poststructuralism in literature, or coming up with a treatment plan for a hospital patient, or designing and building a robot, doing—*designing*—remains the linchpin of an activated education. For the mind to fully, truly grasp something, then the hand must grasp as well—although perhaps not for the reasons one might think.

—

To stick to the world of mechanical engineering, for an example of someone saved by the power of hands-on, complex problem solving, one need look no further than Woodie Flowers himself. Outlandish machinery was a constant in Flowers's life starting at a young age. During his youth in rural Louisiana, local sawmills produced literal tons of waste wood, much of which took the form of giant, live-edge slabs. At most sawmills, two unlucky fellows would have the dubious pleasure of hauling those slabs over to a burning woodpile and chucking them in. "Hot as hell, insufferable job," Flowers said. So his father, a welder, "took a piece of 30-inch-diameter pipe, welded teeth on the outside, spun it at 2,000 RPMs, so it looked like the most evil thing you've ever imagined." He would drop the slabs onto the spinning, toothed cylinder and shoot them off into the distant fire. "Bang!" he said. "You could throw it a hundred yards, no problem." His father's aim was so accurate, "Dad could almost stack stuff with the thing."

In high school, Flowers tried to pick up a date in an army vehicle with wheels in front and tank-style treads in the rear. "I was so happy I was going to be able to pull into the local Dairy Queen in a half-track," he said, but his date refused to get in. His crowning glory was his hotrod roadster, which he built out of found materials. "I didn't have any money, so everything was pretty crude, but it was really fast-accelerating," he said. "I got a ticket one time for no headlights, no taillights, no proper exhaust, a whole list of stuff. No fenders." He chuckled. "I survived."

Despite his budding engineering chops, the young Flowers had no plans to attend college. In his final high school semester, however, a social studies teacher who knew about his interests pulled him aside. He'd noticed that Flowers couldn't fully extend his left arm, the result of a fall out of a tree in the second grade that had fractured it in multiple places. The teacher said, "Boy, we got to get you declared a cripple." An orthopedic surgeon from the nearest hospital made Flowers's disability status official, which equated to a rare opportunity: a college scholarship. "I gave up on getting a job in the oil field and buying a good Corvette," he said. His path took him instead to Northwestern State University at Natchitoches, Louisiana, then Louisiana Tech, then MIT.

When Flowers spun an abbreviated version of this tale at the

2.007 competition opening ceremonies, it sounded familiar to one student in particular: Richard, the dark-horse contender, whose robot had sent concern into the heart of even the unflappable favorite, Tom Frejowski. Richard was a couple of years older than his classmates, which created some distance in terms of temperament. In the lab, while others clustered in groups of two or three, he could often be found working alone, a quiet smile on his face. Still, the first time his robot had fired up its thruster spinner, a crowd immediately gathered, drawn by its Harley-Davidson growl.

If his motor was the only one of its kind in the class, so too was Richard. He'd spent the first part of his childhood in a suburb in southwestern Virginia, where his father worked as a biology and biochemistry college professor. Then, when Richard was fourteen, the family made a change. They moved an hour north, to live and work on a farm. And Richard's homeschool education, which had been at least somewhat structured up to that point, became decidedly less so. The family had moved in the summertime, and that season, Richard began farming in earnest: beef cattle, organic vegetables, organic seed crops. Then in the fall, when it would normally have been time to start a new school year, "It just never really happened, because we were very busy," he said. "My parents, they just basically told me, 'This is a really good educational opportunity. You'll probably learn more here doing this than you would doing schoolwork right now.' "

One of the things he learned about was farm equipment. He found himself responsible for fixing tractors, cars, and ATVs. Along the way, he gained a general sense of where and how machines fail. He also built farm equipment from scratch. Funnily enough, the homemade machine he was proudest of—his equivalent of Flowers's hotrod—was a device that blew air through a column of seeds to remove shells and stalks and rocks. "I built a winnower," he said—but sorting the seed from the chaff proved harder than he'd expected. "I was trying to figure out how to make a bunch of different gates that I could open and close to adjust things," he said, "and it just didn't really work. I wasn't able to get the control I wanted to." Ultimately, he jerry-rigged things until it functioned—a fantastic achievement for a young, untrained engineer—but only inefficiently, and without granting Richard any degree of fine control over the seeds he was cleaning.

Eventually, he realized he wanted a more formal engineering education. Now eighteen years old, he decided to forgo a high school degree and instead studied enough on his own to qualify for an associate's program at a local community college. From there, he applied as a first-year to MIT.

Flowers, upon learning about Richard's background, speculated that MIT's director of admissions, a Course 2.007 alum, might have had something to do with his presence. Whoever made the call to admit him, it was the right one. Richard's robot had turned the heads not just of the other students but also of his professors.

"The VS-11"—a small electric motor provided to all 2.007 students—"is a powerful motor, and he took some gears *out of it,*" explained Amos Winter. "I don't think we've ever seen that."

Of all the motors available to 2.007 students, the VS-11 was capable of producing the most torque, but most students had avoided using it for thruster-spinning purposes because it was a servomotor, designed to move slowly to a set position, and not beyond 180 degrees of rotation. Modifying it to turn continuously was relatively easy—"a hack everybody does," Winter said—but even then, though quite torque-y, it would turn only phlegmatically.

What Richard realized, however, was that the VS-11 only appears to turn slowly on the outside. The *actual* motor, hidden inside the VS-11's plastic casing, spins in a blur. Between that inner motor and the outside world is a series of gears that transform the inner motor's frantic speed into measured, implacable torque. "Yeah, the gear ratio on those is, like—I don't know—200 or 400 to 1 or something. It's enormous," Winter said.

What students normally do, if they want to use the VS-11 to turn a wheel quickly, is build an external gearbox to translate its torque back into speed, which inevitably introduces friction and effectively robs the VS-11 of much of its power. Richard, instead, cracked open the motor's casing. "What he did is he took out some gears, so then it was probably like 10 to 1," Winter said, "and he had very few gear stages and, I think, very good bearings, so it didn't lose a lot of torque from friction." The hack required not just a watchmaker's touch but also the replacement of the motor's circuit board with one Richard had home-cooked for the job.

Winter shook his head in admiration. "That dude is brilliant."

TEACHING PROBLEM SOLVING

By the first match of the day, it was clear that Course 2.007's students had learned plenty of *things.* At the start of the semester, Amy Fang had come in with a decent theoretical grounding in how gears affected the torque output of a motor, for instance, but she didn't know how to fit gears together to form a transmission. By the end of the class, she'd won an award for the transmission she'd built, which transferred the power of not one but three motors to the X-wing's thruster.

A harder question was whether students had improved their ability to solve complex problems, a trait that some educators consider unimprovable. Brandon, after aborting his first plan, had thrown together a serviceable, thruster-spinning robot in the space of a handful of weeks. Then he did it again: He built a second robot for his wingman Josh to pilot, this time in a matter of *days.* This auxiliary robot, though legal according to the rules of the contest, had to be built with a dramatically different control scheme, governed by an Xbox videogame controller rather than a model airplane remote, which posed an entirely new engineering challenge. Nevertheless, his turnaround time was remarkably short. Somewhere along the line, he had clearly attained some hard-to-pinpoint knowledge about engineering and design.

What was the identity of that knowledge, exactly? Fascinatingly, both Sanjoy Mahajan and Mitch Resnick—representing inside-out and outside-in educational traditions, respectively—laid claim to aspects of the learning being done in Course 2.007.

Resnick's outside-in claim made the most intuitive sense. By literally building things, students caused their schematic trees to grow markedly fuller: flowering oaks where there had once been mere saplings. "As soon as I start creating, that gives me new ideas," Resnick said. For instance, when a student learns to use, say, the automatic lathe workstation in the Pappalardo Laboratory to fabricate a grooved shaft of some sort, other possibilities posed by the lathe begin to unfold, and soon that same student finds herself using it to create, say, flanged bushings to add rigidity to the joints of her scissor lift. If merely learning to use a machine complicates one's reality, learning to *build* one does so on a whole higher order of magnitude. For instance, everyone graduates from Course 2.007 able to invent robotic solutions to everyday annoyances like household chores. Their immediate surroundings become solvable. The world, in a small way, becomes changeable.

But still, the question of whether the course improves deep-seated problem-solving skills, or merely adds to the highly organized, interconnected branches of data housed in students' long-term memory, remains open. After all, knowledge of what a lathe does is just that: raw knowledge. In fact, a good deal of Course 2.007 consisted of the direct instruction of principles—often delivered through the media of online tutors and videos. And, as Mahajan pointed out, although there was plenty of discovery-style learning going on in 2.007 at a surface level, a closer look revealed the course to be highly *scaffolded:* organized to break leaps of discovery into more manageable steps. Students like Brandon and Amy Fang and Z may have felt like they were chucked into the deep end of the learning pool and abandoned to sink or swim, but in point of fact, the course strategically placed a sequence of life preservers within reach. These included the course's series of "physical homeworks," which forced students to achieve certain engineering milestones before starting in on their competition bots. "They're not thrown in. They're not just told, 'Okay, here, go figure out how to build a robot and try experiments,' " said Mahajan. "They're taught theory. They're taught how to work together. And so that's not discovery learning at all."

—

To mark the official start of competition, the Chorollaries, an MIT a cappella group, sang the National Anthem, which was greeted with hearty applause. Then, blasting forth from loudspeakers, came another anthem chosen to lift the heart of every geek in attendance: John Williams's *Star Wars* theme. The stage set, Winter and his co-lead, Sangbae Kim, burst through the curtains clad, respectively, in startlingly realistic Darth Vader and Chewbacca costumes. The crowd erupted.

The opening match was Richard's first contest of the night. His robot, named Tornado, trundled autonomously up to the lower thruster and spun it at high speed, emitting its characteristic growl. It worked by means of a friction drive: turning a small wheel against the face of the thruster, like a tiny unicyclist riding on a record turntable—a design strategy that, of the final thirty-two competitors, only Richard and Amy Fang had attempted. Then it did what Tom's robot couldn't. Now under Richard's manual control, it flew up the elevator, repeated its feat on the upper thruster, and, to put a cherry on top, slammed a button that caused the music from *Star Wars*' Mos Eisley Cantina to play, which added another smattering of points to his total. By the end of the round, 937.5 points glowed on the scoreboard, the competition's highest score yet. His opponent never had a chance.

The rest of the first round of competition proved as ruthless a winnower as any Richard—or E. L. Thorndike—had ever designed. For the most part, a wide variety of lovingly crafted robots fell before an onslaught of dedicated thruster-spinners—clearly, this year's winning strategy. Brandon and his roommate, Josh, each had a pair of these spinner bots, and each served as the other's co-pilot in the competition. Their first team-up—Brandon's round, with Josh assisting—proved sweat-inducing, with the robot under Josh's control never quite aligning with the lower thruster. Brandon, however, pulled out a win with mere seconds remaining by climbing the elevator and spinning the upper thruster. "That's gonna be awkward when they get home tonight," Winter deadpanned over the loudspeaker.

In their second team-up—technically Josh's match this time—the duo made hasty work of Amy Fang and her beloved Dodocopter. While approaching the lower thruster, Dodocopter somehow put one wheel in the trench running down the middle of the game board, beneath the X-wing's fuselage. Amy tried to spin the thruster before the inevitable took place, but to no avail: The robot teetered and fell. The Dodocopter was extinct, and Winter was sad to see it go. "Amy did some killer analysis," he told the audience. "I love her design."

That same central trench would swallow Josh's chances as well, in the sweet sixteen. By working together on both of their bids, Brandon and Josh had essentially given themselves two shots at competition glory. Now they were down to just one—Brandon's bid—and the odds seemed not in his favor. His round-of-16 match was against James Li, who, like Brandon, had a partner to help him drive his two robots, named Bonnie and Clyde. The bigger of the two, Bonnie—"Just like the praying mantis, the female is larger," intoned Winter—extended a telescoping tower up to the *front* of the top thruster. Brandon raced up the elevator and spun his top thruster to its maximum speed of 25 radians per second. Li's Bonnie, meanwhile, fighting some sort of alignment problem, achieved only a fraction of that.

Still, a fraction could go a long way, because by now the smaller Clyde, too, had reached the upper deck and inserted a hook behind the dangling, weighted base of the lightsaber situated in the center, right above the X-wing's cockpit. The lightsaber's heavy handle acted like a pendulum, and any competitor who pulled it far enough could multiply her point total by as much as three. Brandon, eyes agleam, began to ram his robot into Clyde. It was the night's first instance of sabotage, which was permitted only for robots that had reached the upper deck, where the two combatants were now entangled. "Wow, that is a Dark Side move!" said Winter. Still, despite Brandon's worst efforts, the lightsaber continued to tip, reaching 45 degrees. It looked for a moment like the match could go to anyone—except for the fact that Josh had quietly been working in the background, spinning the lower thruster. Ultimately, it wasn't Brandon's sabotage so much as Josh's quiet success that carried the duo to the round of eight.

Z, meanwhile, who had worked sabotage into his plan weeks ago, found himself dreaming up new schemes for each opponent. His larger robot could spin both thrusters from the front, like several others, by extending its thruster-spinning fork up on an accordion-style platform. But his most cunning stroke of genius was the realization that this rising platform could also serve as his own personal elevator, capable of depositing a smaller, sabotage-ready bot on the upper deck right at the start of the round, which could permit all manner of mischief.

Looking at his tournament bracket, he saw that he would first run into trouble in the form of Patrick Shin, whose robot could spin both thrusters from the front. "Honestly, I may have my second driver drop down to block him," he said—a maneuver that Z referred to as "going 'Fast and Furious.'" It would be technically legal, so long as it didn't endanger any nearby humans.

Even Alex Hattori, the student who had competed on the television program *Battlebots*, sounded concerned. "I'm glad I'm in the other bracket," he said.

In practice, however, the transfer of the smaller robot from Z's homemade elevator to the X-wing's upper surfaces proved trickier than he had anticipated, necessitating a hastily constructed gangplank. In his first two rounds, the small sabotage bot fell from this bridge both times, useless. The second time around, the crowd, which now understood Z's strategy, groaned at the sight of the robot plummeting; they'd wanted to see it do its dirty work. Still, by the end of the second match, the crowd was chanting "Z." He had made it to the quarterfinals. "I never thought I would be in the eight," he said, eyes bulging. He embraced his co-pilot, Gabriel, and then both raced away to tinker with their robots. In fact, he'd make it even further, winning the next round handily against a bot that oscillated free of the top thruster at precisely the wrong moment. Z would move on to the semifinals, where Richard waited with his Tornado of doom.

Brandon, meanwhile, was up against Alex Hattori of BattleBots fame, whose twin robots were theoretically capable of racking up astronomical sums. Instead, however, the match unfolded in a cavalcade of errors on both sides—including a moment when Brandon's robot fell on top of the one piloted by Josh—eliciting frustrated moans from the crowd. It looked like a tie was in the making, in which event the lighter pair of robots would advance. But Josh and Brandon had one advantage Alex lacked: they had read the fine print in the rulebook, which permitted them to lay down a single stormtrooper on their side of the game board at the beginning of the round. If pushed into the central trench, it could earn them a measly five points—essentially a concession built into the competition for students with the simplest robots, all of whom had been eliminated yesterday. Now, with about ten seconds left, that loophole was their only hope. Josh's robot was still upright, but it wasn't really built for the job, and the moment it drove up against the heavy, metal action figure, it seemed to stall. "Five," Winter announced. The stormtrooper started to slide, and the crowd, watching the action on jumbo screens, began to murmur. "Four." Josh pushed the throttle all the way up. With three seconds left on the clock, his robot plummeted into the trench, dragging the stormtrooper with it. Brandon's team had won by five points. And suddenly, the audience, which had been cheering throughout the night in a good-natured yet contained sort of way, lost its collective mind. Many had come merely to support their classmates out of friendly obligation, but now it was becoming clear that they were watching the best sporting event of the year.

—

Part of why the crowd in Johnson Rink reacted so passionately to Brandon and Z was that they had begun to exude the sort of confidence that one loves to see in professional competitors. There was a moment in the quarterfinals when, with the clock still running, Z turned and high-fived his partner, which called to mind the sprinter Usain Bolt's tendency to glance back at his cloud of dusty competitors in the 100-meter dash. The self-assurance Z displayed was perhaps a little cocky, but it was also captivating to witness in someone who had only recently achieved a degree of proficiency.

Woodie Flowers later alluded to this feeling when he described what was to him the secret purpose of the hands-on, *Manus* part of an MIT education. Education is what happens when you learn "to think using calculus," he reiterated—but that's actually a two-part proposition. The first part is outward-facing: You have to understand how calculus relates to the world around you at a deep level, so that calculus becomes a tool applicable to a wide variety of situations. It must be broadly contextualized and also, in the wording favored by Sanjoy Mahajan, "overlearned," so you can reference it on the go, without clogging up your working memory.

The second, inward-facing part is less intuitive but just as important. Education means gaining a second-nature understanding of how *you*, as a potential agent of change, might use your knowledge and skills to affect the world. "I believe that true education is about a process that allows one to develop rational self-esteem," Flowers said. That is, for anyone hoping to affect the world, prodigious knowledge and skills are never enough; you must also prove to *yourself* your mastery of that knowledge, of those skills, in order to understand their wider relevance.

"I used to go to conferences," talk about 2.007, and then "get really frustrated," Flowers said. "A faculty member from another school would come up and say, 'Yeah, we do creative exercises. I had this consulting job, and I had a problem that I had no idea how to solve, so I gave it to the students as a creative exercise.' And you just want to deck them. That's the dumbest goddamn thing you can ever imagine doing. I mean, why would you take somebody's creative ego and trash it from the beginning?"

Nurturing a creative ego, rather, is where the *Manus* aspect of education excels. At the beginning of the semester, Winter had introduced the design process he and the other 2.007 lecturers would propound throughout the course. "Design doesn't work like, 'you have a great idea, you make a Saturn V rocket, and you go to the Moon,'" Winter had said. "Try to picture a Saturn V rocket in your head. Not just the shape, but, like, one of the O-rings in one of the thrusters in the bottom." The rocket was sixty feet taller than the Statue of Liberty, filled end to end with complex machinery and electronics. "It's massive, massive, massively complicated. And so people just don't have the cognitive ability to carry all of these details in their head. They have to break it down to smaller, more tractable parts."

He loaded a slide of a diagram that was originally created by the product designer Damien Newman. A black squiggle on a white background, it looked like someone's chaotic pen-and-ink signature, except that from left to right its frenzied loops hewed toward some invisible central axis, until eventually a single flat horizontal line emerged. "It basically starts out with research and conceptualization," Winter said, gesturing to the squiggliest, leftmost part of the diagram, whose multifarious loops represented the many routes that might lead to a complex problem's solution. Research and hard thinking eliminate a number of these possibilities. Next, he gestured at the slightly more orderly middle section of the diagram. "You evaluate your ideas, you narrow 'em down, you get more refined, and eventually"—now pointing to the spot on the diagram where the several remaining lines coalesced into one—"*boop*, you come up with a final design."

When first approaching a complex design problem, "you can believe six impossible things before breakfast. Near the end, you're making an absolute prediction about what the universe will do," said Flowers. "I know really good designers who are comfortable everywhere in the space, and they never get confused about where they are," he said. "To me, elegant design is people that run back and forth"—to different parts of the design squiggle—"with fluidity and precision."

Fluidity and precision: If that sounds like the sort of thing that draws spectators to sporting events, it should. And indeed, if you managed to chemically isolate what the crowd in Johnson Rink was applauding when they stood up and stamped their feet for Brandon and chanted Z's name, it would have been exactly that: the precision on display in their robots and the fluidity apparent in their ability to wield them.

In the fluidity and precision of the Course 2.007 competitors, too, I see the greatest argument for further disentangling the taken-for-granted aspects of our inherited educational structures. In 2.007's pocket universe, not only are students freed from fear, but intractable points of pedagogical disagreement, such as the teachability of problem solving, begin to feel less pressing. Unlike "problem-solving skills," which Z and Brandon might or might not have had in spades from birth, their self-confidence as roboticists was something they had certainly gained as a result of 2.007's deliberate, highly scaffolded approach to hands-on instruction. Once you accept that nurturing students' creative ego is essential, then ancient, Jesuitical disagreements about problem solving simply begin to lose their urgency. Regardless of whether such skills can be taught, you're going to have to use hands-on, scaffolded, discovery-style pedagogical tactics for the development of a creative ego anyway. Only the *Manus* part of *Mens et Manus* can foster the sense of self-confidence you need to survive the ups and downs of the design squiggle: a sense of self that will keep you anchored whether you're designing a government policy, a piece of writing, a robot, or a symphony.

—

Or even if you're designing new educational standards. Looking back, without precisely such a sense of creative ego, I could never have even conceived of a viable pathway for the MicroMasters. The creation of new standards in any complex field is a bit like building a rocket, at least in the sense that there are far too many pieces (and interested parties) to ever account for all at once. When we set out to establish a new RFID standard for the world's supply chains, I lucked into a winning formula (and team of collaborators) for aligning all the moving parts. One consequence of that experience was that, when it came time to plan out the MicroMasters, I knew it could be done. I understood what the points of uncertainty would be like and what it would feel like to overcome them. And I also knew that if we found a way to somehow attach value to the standard, we could make it stick. And that's precisely what we did: We thought very intentionally about what makes educational credentials valuable to their owners, and we created a version that married maximal value with minimal exclusivity.

That's a formula we'll hold on to moving forward. Ultimately, if existing institutional structures continue to stand in the way of widely available, cognitively user-friendly learning, then perhaps tomorrow's learners will need to hitch themselves to a new kind of rocket. As we've seen, a whole host of organizations are experimenting with ambitious new schemes for how students might move through curricula. At MIT in particular, there are enough of these now up and running or in the works that, when put together, they begin to look almost like a coherent, alternative pathway for advanced learning.

And indeed, like a Saturn V rocket, there are lots of separate parts to keep track of. Here are the most critical modules. Perhaps the most mature element in this alternative pathway is our ever-growing complement of free (or nearly free) online courses, which, as we've seen in the MicroMasters, can be used not just to teach, but also as admissions criteria for further levels of instruction. Such further instruction, meanwhile, might include not just master's degrees, but also intensive midcareer bootcamps, which are now running continuously at MIT. These cram an entire semester-long course's worth of hands-on learning into a single week—admittedly suboptimal from a spacing perspective, but necessary for our bootcamps to fit into the packed schedules of the attendees. (You can often tell where a bootcamp is taking place by the piles of exhausted thirty-somethings catnapping on the lawn.) For the level beyond that, we're setting up a system of one-on-one apprenticeships, not with professors as advisors so much as "entrepreneurs, investors, corporate executives," explained Erdin Beshimov, who is leading the program.

Just these three ingredients might themselves be enough to create a viable, nontraditional route to intellectual superpowers in a given field. First, online learning would deliver hard knowledge and skills, then hands-on bootcamps would contextualize that knowledge while nurturing a healthy creative ego, and finally an apprenticeship would offer the chance to turn around and apply that knowledge directly to a field's emerging problems.

This potential progression would necessarily exist apart from graduate education as we know it, and be undergone mainly by people for whom traditional advanced degrees aren't quite the right fit. But there's no reason that traditional degree programs, too, shouldn't be armed with some of the same sorts of instructional ingredients. At the college level, blended or flipped learning is one obvious, cognitively user-friendly approach that fits easily into traditional educational infrastructure. (In fact, as my colleagues remind me when I get too overexcited about these things, humanities professors have always run flipped classrooms, simply by asking their students to "please do the reading before class.") One less obvious opportunity raised by such a setup, however, is that the online part of a flipped course need not be home-cooked at the same college where it's being taught. To encourage other colleges and universities to incorporate our online materials into their flipped courses, the physicist Krishna Rajagopal, MIT's Dean for Digital Learning, has launched a program called the xMinor. The general idea is that if you're a student at perhaps a small liberal arts or community college and you want to take a course that's not offered—say, quantum computing—you could take that course remotely from MIT (or whoever was offering it). You would still have a local professor, however, who would make those online materials relevant to your larger education at your college—that is, who would provide the *manus* to the online component's *mens*. "The role of the educator is just as central as it was before," explained Rajagopal. In this sort of blended course, "the educator is doing the blending."

In order to combine all of these loosely connected efforts into something coherent, I've been giving particular thought to the one educational standard that could rule them all: a modular, distributed transcript, owned by students, not their educational institutions. A universally recognized network of unfakeable transcript entries, not too different from the generalized network used to track the billions of different RFID tags swirling around the world, would permit a far more free-flowing experience for students, enabling them to sample from multiple institutions of higher education, to mix and match on-campus with online courses, and to earn traditional degrees as well as newfangled credentials as needed. Such a system need not be limited to the stuff of transcripts as we know them—namely, numerical grades. They could also include portfolios of projects, granting admissions offices a fuller view of their applicants than is currently possible.

We're still in the early days of imagining such a system, with a long line of obstacles both known and unknown in front of us. Our experience with the MicroMasters has proved, however, that the creation of new educational standards is far from impossible. We're like 2.007's freshly minted roboticists: surrounded by a world of problems that once seemed out of reach, but now appear solvable.

Even if my wildest fantasies of success come to pass, however, and far more knowledge soon becomes far more accessible to far more people, there will still always be schools that traffic in exclusivity, conferring name-brand value on attendees via the coins of "who you meet" and "the fact you got in." But crucially, top-notch learning will no longer be locked behind such doors. Increasingly, in the years to come, the "stuff you learn" will be available to everyone.

THE SHOWDOWN

The Course 2.007 semifinals pitted the upstarts from Winter's lab section against the overall favorites: Z versus Richard, Brandon versus Tom. For once, the outcomes of matches involving Richard and Tom seemed not to be foregone conclusions, once you factored in both Z's and Brandon's now-apparent lust for sabotage. Z immediately fell victim to mechanical failure, however, and although Brandon managed to drop his robot from the X-wing's fuselage onto Tom's, he struck only a glancing blow—too little, too late.

In the final showdown between Richard and Tom, Black Sabbath's "Iron Man" was blasting throughout the arena. Tom crouched down, eyeballing his robot's trajectory. The winner was clear mere seconds into the match, when Richard's Tornado struck its target slightly left of center. Its thruster-spinning wheel began to turn, but the thruster remained immobile, and Tornado slid off to the side, losing contact. And that was it: Richard would score no points in the autonomous period, and although he quickly assumed manual control and earned points on both thrusters, he couldn't overcome Tom's lead. Tom, the favorite from the start, carried the night, and Winter and Sangbae Kim carried *him:* on their shoulders, back and forth in front of the crowd.

But there was one other match that took place before the night ended—one that proved more of a crowd-pleaser than even the finals. It was the consolation match between the remaining semifinalists, and, in effect, the final competition for champion of Winter's lab section. Both Brandon and Z were prepared to do whatever was necessary to win.

By the end of the autonomous period, Z's double robots had pulled up to the starfighter's fuselage and waited at the ready. The manual control period began, and Brandon raced up to the elevator. Z's gangplank unfolded and now Z's second robot, controlled by Gabriel, rolled onto the top of the X-wing, beating Brandon to the upper deck. Z's personal elevator had finally functioned as intended and the audience, witnessing this long-anticipated moment, came to life. The decibel level only increased when Brandon too arrived on the top deck. Gabriel zipped over to prevent him from coupling with the top thruster. There was a moment of confusion; the action became hard to follow, and then the top thruster began to spin. Strangled yells could be heard from the audience, and then a video feed showed what was happening: Gabriel was trying to dislodge Brandon from the top thruster while Brandon, locked in, was driving his wheels with the full might of his electric motor. The thruster reached its maximum speed of rotation: points to Brandon. But then it looked like the tables might turn just as rapidly. Josh had spun the lower thruster on Brandon's side to maximum speed, as had Z to the top thruster on his side. Now all that remained was for Z to shrink his scissor lift down and spin the bottom thruster. If Brandon could somehow stop that, the match would be his. Gabriel's robot was still in his way on the upper deck, but he was no match for Brandon as a driver, and Brandon skirted around him and drove out alone on the fuselage of the X-wing. The spectators, sensing what was about to happen, literally rose to their feet. Z plugged into the lower thruster and began spinning it—how fast, no one knew—while Brandon teetered above him. Then, with a resounding crash, Brandon landed on the flat top of Z's scissor-lift robot and collapsed it to the ground. When the dust settled, the scoreboard registered an improbable tie: 312.5 to 312.5.

And so out came lab instructor Danny Braunstein, solemnly hooded in a brown Jedi robe, with a set of homemade scales. Brandon placed both of his robots on one side; Z placed his on the other. For a long moment, the scales remained locked, level. Then Braunstein pulled a lever, freeing them with an audible click. One set of robots fell toward the floor and the other set rose in triumph, held aloft like the hand of a victorious prizefighter.

It belonged to Brandon.

In the immortal words of Darth Vader, the learner was now the master.

Epilogue

In 2019, two years after he'd delivered his address to Brandon and the rest of the Course 2.007 students, Woodie Flowers died following surgical complications. He was seventy-five. The news came as a blow to MIT, and to a far wider community of learners as well. In addition to the alumni of his now-famous MIT course, Flowers also reached millions through FIRST Robotics, the globe-spanning organization he helped found, which brought robotics competitions to students in dozens of countries. Factor in the three years he spent as host of the PBS television show *Scientific American Frontiers*, and the vast number of high school and college courses built on 2.007's model, and it's no exaggeration to say that he contributed to the education of a respectable chunk of the world's population. Already, his influence is of the same order as that of some of the most important names in the educational pantheon.

Two hundred years earlier, the sound of cannon fire heralded the end of the legacies of two members of that pantheon: Andrew Bell and Joseph Lancaster. In 1817, while Lancaster's schools were still popping up like mushrooms in the young United States, DeWitt Clinton, New York's governor and Lancaster's most prominent proponent, announced the start of construction on the Erie Canal. The ambitious waterway would connect the Great Lakes to the Atlantic via the Hudson River and its muddy tributary, the Mohawk. "Clinton's folly," as the 363-mile project was known, was originally derided as the impossible scheme of a grandiose politician, but it was finished under budget in the span of eight years. In 1825, the State of New York announced its completion with a sequential cannonade running the length of the waterway, starting on the shore of Lake Erie and finishing in New York City. The gunfire took an hour and a half to travel from lake to shining sea.

The canal was eminently worthy of celebration: It cut transportation costs of goods by 95 percent and turned New York City into the primary point of commerce for not just the northeastern United States, but also the upper Midwest and parts of southern Canada. It also utterly upended life within its reach. Prior to the Erie Canal, farmers and artisans had only their neighbors to compete with. But once that cannonade fired, there was something new to consider: the national price of beets, of boots, of barrels of whiskey. "Every wheat farmer was suddenly competing with every other wheat farmer across New York State and beyond," writes the education historian David Labaree in his book *Someone Has to Fail*. In a few short years, a host of small, local businesses, especially family workshops, became economically unviable, no longer able to compete on price with mass-produced goods from faraway cities. As canals, turnpikes, and railroads spread, once-reliable ladders to middle-class prosperity disappeared, and so did much of the social glue that had held agrarian communities together. The public responded in part by embracing mass education: expanding its purview from Lancaster's needy cases to just about everyone. In the process, its framers set up educational norms that remain with us today.

Something very similar happened in response to the Industrial Revolution of the late nineteenth century, as mass production became mechanized and stand-alone factories gave way to larger corporations. Once again, tried-and-true ladders to the middle class disappeared; once again, a sense of social upheaval led to the frantic construction of institutions. By now, education was becoming the clearest path to the middle class for many families. They began clamoring for high school diplomas for their children, and the then-ascendant administrative progressives decided it was urgent that someone weigh and sort these students. As we've seen, their methods were less impartial than they supposed, and we still live with many of the structures and practices they laid down—including the educational winnower.

Today, we're experiencing a moment remarkably similar to those earlier watershed periods. Income mobility, as Flowers pointed out in his address, has stagnated. Automation in the workplace is claiming certain types of jobs while opening up new ones, a remarkable parallel to what happened over the course of the nineteenth century, when mass-production-by-hand gave way to mass-production-by-machine. The regional market revolution of the early nineteenth century, too, is a notable antecedent for today's global flow of jobs, people, and goods (the last of which I suppose I've contributed to as much as anyone, considering RFID's role in greasing the wheels of global trade). As the cliché goes, trade helps more people than it harms, but it does create winners and losers, and it can move once-reliable economic ladders out of sight when you're not looking.

These conditions, combined with the rapid pace of technological change, demand an education that is simultaneously able to keep up with the times, yet also timeless. As Max Ventilla put it, the aphorism "Give a man a fish, he'll eat for a day; teach a man to fish, he'll eat for a lifetime" makes a degree of sense—"but that's assuming that the way that you teach him how to fish stays relevant for the rest of his life." Even as the specific rules of how to earn a living change, however, broad skills never go stale. To the list of such classics as "learning to learn" I'd add comfort in the face of complexity, control over one's own working memory, and, perhaps most important, to use Flowers's term, the development of a creative ego. That doesn't mean we should give up on direct, subject-specific training, but such training must become more forgiving in terms of learners' time, location, and money, and more agile in the face of change. What we need is a patchwork approach—a *mens et manus* approach—to give students both the hard facts they'll need to understand the world and the hands-on skills they'll need to wield them: to *think using calculus*. The MicroMasters to master's progression is an example of a two-part patchwork quilt, but it's far from the only possible model.

If we do soon experience yet another major educational reshuffling, the timing couldn't be better, because it might give us the chance to root out and replace the antiquated notions still influencing how we're expected to teach and learn. Thanks to the unceasing work being done up and down the high-rise of the cognitive science disciplines, we now have the capacity to structure instruction that doesn't interfere with the biological and psychological processes underlying learning, but rather supports them. And thanks to new, often (but not always) tech-enhanced tools and methods, we can take such approaches and make them both more widely accessible and more flexible, tolerant of the differences that make us unique.

Perhaps most crucially: Only today do we have the advantage of hindsight. We may finally have the historical perspective we need to avoid the mistakes of prior generations of educationalists. As William James wrote in 1899, "Psychology is a science, and teaching is an art; and sciences never generate arts directly out of themselves. An intermediary inventive mind must make the application." We must think about learning not only as scientists, but also as appliers of that science: designers, engineers, artists, teachers, *learners*. By so doing—with a modicum of decisiveness and a minimum, I hope, of hubris—we can improve lives in a tangible way, and direct more of the world's latent potential toward the hard problems we face.

For too long, we've blindly obeyed educational traditions built on precarious scientific evidence, and consequently impeded learning everywhere. Now, the opportunities presented by a more intentional approach—to place learning above winnowing; to place access above exclusivity—are too promising to ignore. There is no better day than today to announce a new age of learning. Before the hour grows any later, let's fire off a cannonade not of gunpowder but of action potentials and get started.

Acknowledgments

This book could not exist without the community of people who urged it and its creators forward. Special acknowledgment is due to the late, prolific, fantastically talented science writer William Rosen. Bill and Sanjay had intended to write a book on learning together, but never got the chance. The world is poorer without Bill, but at least it still has his books in it—a small but real solace.

Yaniv Soha, our editor at Doubleday, and Eric Lupfer, our agent, provided constant insight and steady hands throughout this book's writing, especially at times when the subject seemed too unwieldy to fit into a single volume. Both deserve our deepest thanks. We'd also like to thank Cara Reilly and the copyediting team who worked on the manuscript.

At MIT, David Shrier helped get this project going in the first place, and Lisa Schwalli, Marisol Tarbares, Melissa Manolis, Marine Brown, and Emma Crist helped a great deal throughout the process. Special acknowledgment goes to Laura White, executive assistant to Sanjay, who managed to thread various needles with grace and competence.

Certainly, this book could not exist in its current form without the vision, support, and hard work that has made MIT's Office of Open Learning possible, provided most importantly by Rafael Reif, Anant Agarwal, Ike Chuang, Krishna Rajagopal, Erdin Beshimov, Community Jameel, and many others. Just as noteworthy are the keen, forward-thinking teams powering Open Learning, who are constantly bringing new educational possibilities to light.

Both within and beyond Open Learning, there is a host of luminaries from the MIT community we'd like to thank individually, including Dheeraj Roy, Nancy Kanwisher, John Gabrieli, Kana Okano, Mitchel Resnick, Sanjoy Mahajan, Mikael Lundqvist, Laura Schulz, Amos Winter, Sangbae Kim, Daniel Braunstein, the whole 2017 2.007 class (especially the semifinalists Brandon McKenzie, Z Liang, Richard Moyer, and Tom Frejowski) as well as the Pappalardo Lab instructors, Karen Willcox, Philip Lippel, Hal Abelson, James Donald, Chris Boebel, Dana Doyle, Battushig Myanganbayar, Fox Harrell, Justin Reich, Philip Altbach, Vincent Quan, Sophie Shank, Susan Silbey, Marc Aidinoff, Tom Smith, Yossi Sheffi, Chris Caplice, Eva Ponce, Srideepti Kidambi, and Paulina Gisbrecht.

Special thanks goes to Joe Coughlin and the multitalented team at the MIT AgeLab, who supported Luke during what turned out to be a protracted writing process.

Outside MIT, our indebtedness extends unabated. In many ways this book got its genesis when Sanjay read *Make It Stick*, by Peter C. Brown, Henry L. Roediger III, and Mark A. McDaniel. Other figures whose writing proved especially crucial include the education historian David Labaree as well as Geraldine Jonçich Clifford, whose 1968 biography of E. L. Thorndike should be considered a classic of both biographic and science writing. Figures who helped in more concrete ways include Eric Kandel, Wayne Sossin, Toshihide Hige, Nadine Gaab, Jacqueline Gottlieb, Robert and Elizabeth Bjork, Louis Schulze, Peter Dourmashkin and the MIT TEAL team, Catherine Nunziata, Claire Wang and her family, Gaetan Juvin, Josh Trujillo, René Ramirez, Josh Dahn and the teachers at Ad Astra, Alison Scholes, Sep Kamvar, Ted Quinn, Max Ventilla, and Alex Kaplan. Our sincere thanks to everyone.

An extra dose of appreciation goes out to people who provided critical insights for this book, some of whom read sections or entire manuscript drafts. In addition to several of those named above, we would like to thank Yarden Katz, Florence Bouhali, Michael Horn, Julia Freeland Fisher, Sherman Dorn, Adam Laats, and Ray Bub.

This book was also built on a network of late-night conversations as well as spare bedrooms and cars. Special thanks goes to Phil McKenna, Philip Picotte, Connor Williams, Daniel A. Gross, Rachel Ann Cole and Justin Watts, Elizabeth and Paul Kelley-Sohn, Ellen and Geoff Stearns, Audrey and Dave Campbell, Ryder Musselman, Jeannie Catmull, Brad Hutchinson, and the Happy Phinney Group.

So many contributed to this effort in big ways and small, in fact, that it's very possible we've overlooked people who lent a hand at key moments or provided a needed bit of insight. We apologize to everyone we've left out.

For their help (and for putting up with us) during this project, we would also like to extend our most sincere thanks to our families, including our parents, Gitanjali, Tara, and Kelsey. Extra thanks to Susan and Eileen for providing childcare for Hope, and to Hope for being born!

We'd like to end on one last note of appreciation for Woodie Flowers.

—Sanjay Sarma and Luke Yoquinto

Notes

INTRODUCTION: THE ADVENTURE BEGINS

x Together with about 70,000: Shreeharsh Kelkar, "The Elite's Last Stand: Negotiating Toughness and Fairness in the IIT-JEE, 1990–2005" (working paper, 2013); http://web.mit.edu/skelkar/www/shreeharsh-kelkar_files/Kelkar-IIT_JEE_ver4.pdf.

xi Saint Jean-Baptiste de La Salle: Clandinin and Husu, *The SAGE Handbook of Research on Teacher Education*, 55.

xii "as common in the classroom": Cuban, *Teachers and Machines*, 19.

xii "the 21-inch classroom": Larry Cuban, "Techno-Reformers and Classroom Teachers," *Education Week* 16, no. 6 (1996): 37–38.

xii In 1961, *Popular Science* predicted: C. P. Gilmore, "Teaching Machines: Do They or Don't They," *Popular Science* 181 (1962): 57–62.

xii "a significant part of every child's life": Papert, *Mindstorms*, 18.

xii "the computer will blow up the school": Seymour Papert, "Trying to Predict the Future," *Popular Computing* 3, no. 13 (1984): 30–44.

xiii "the whole idea": Toffler, *Future Shock*, 355.

xiii "students in batches": Michael B. Horn and Meg Evans, "A Factory Model for Schools No Longer Works," *Milwaukee Journal Sentinel*, June 29, 2013.

xiii "tends to alienate teachers": Greg Anrig, "Why the New Teachers' Contract Is Great News for NYC's Students," *The Century Foundation*, June 3, 2014; https://tcf.org/content/commentary/why-the-new-teachers-contract-is-great-news-for-nycs-students/.

xiii have echoed the metaphor: Linda Darling-Hammond, "To Close the Achievement Gap, We Need to Close the Teaching Gap," *HuffPost*, June 30, 2014.

xiii "a kind of halfway house": Gatto, *Underground History of American Education*, 168.

xiii "There is no need": Salman Khan, "The Founder of Khan Academy on How to Blend the Virtual with the Physical," *Scientific American*, August 1, 2013.

xiii For one thing: Audrey Watters, "The Invented History of 'The Factory Model of Education,'" *Hack Education*, April 25, 2015.

xiv elite colleges in particular: Jeffrey Selingo, "The Two Most Important College-Admissions Criteria Now Mean Less," *The Atlantic*, May 25, 2018.

xv "probably the vast majority": Caroline M. Hoxby and Christopher Avery, "The Missing 'One-offs': The Hidden Supply of High-Achieving, Low Income Students," National Bureau of Economic Research, No. 18586 (2012).

xvi Since the early 1980s: Thomas D. Snyder, Cristobal de Brey, and Sally A. Dillow, "Digest of Education Statistics 2017, NCES 2018-070," National Center for Education Statistics (2019); "Trends in College Pricing 2019," College Board, November 2019, https://research.collegeboard.org/trends/college-pricing/resource-library; Emma Kerr, "10 Most, Least Expensive Private Colleges," *U.S. News & World Report*, September 9, 2019.

xvi There are a number of reasons: Michael Mitchell, Michael Leachman, Kathleen Masterson, and Samantha Waxman, "Unkept Promises: State Cuts to Higher Education Threaten Access and Equity," Center on Budget and Policy Priorities, October 4, 2018.

xvi As the MIT economist David Autor: David Autor, "Skills, Education, and the Rise of Earnings Inequality Among the 'Other 99 Percent,'" *Science* 344, no. 6186 (2014): 843–51.

xvi Indeed, one intriguing, model-based analysis: Aaron Hedlund and Grey Gordon, "Accounting for Tuition Increases at US Colleges," *2017 Meeting Papers*, no. 1550, Society for Economic Dynamics, 2017.

xvi Meanwhile, at lower-tier colleges: Paul Fain, "College Enrollment Declines Continue," *Inside Higher Ed*, May 30, 2019.

xvi The relationship between family income and college attendance: Raj Chetty, Nathaniel Hendren, Patrick Kline, and Emmanuel Saez, "Where Is the Land of Opportunity? The Geography of Intergenerational Mobility in the United States," *Quarterly Journal of Economics* 129, no. 4 (2014): 1553–623.

xvii Although in raw terms: Raj Chetty, John N. Friedman, Emmanuel Saez, Nicholas Turner, and Danny Yagan, "Mobility Report Cards:

The Role of Colleges in Intergenerational Mobility," NBER Working Paper no. 23618, National Bureau of Economic Research (2017): 35.

xvii elite schools have the edge: Stephen Burd, "Even at Private Colleges, Low-Income Students Tend to Go to the Poorest Schools," *New America*, May 18, 2017.

xvii "Ironically," Stanford's Hoxby has said: Caroline Hoxby, "Students of Color Don't Apply to Top Schools, but They Should," *Tell Me More*, NPR News, January 9, 2014; Amanda Ripley, "Why Is College in America So Expensive?" *The Atlantic*, September 11, 2018.

xvii "the dull remain dull": Ravitch, *Left Back*, 160–61.

xviii Soon, however, his test was coopted: Nancy Beadie et al., "Gateways to the West, Part 2: Education and the Making of Race, Place, and Culture in the West," *History of Education Quarterly* 57, no. 1 (2017): 94–126.

xviii considered it a given: For a discussion of Terman's willingness to venture beyond the available data, see Russell T. Warne, "An Evaluation (and Vindication?) of Lewis Terman: What the Father of Gifted Education Can Teach the 21st Century," *Gifted Child Quarterly* 63, no. 1 (2019): 3–21.

xviii "chiefly a matter": Lewis M. Terman, "Were We Born That Way?" *World's Work* 44, no. 660 (1922): 659.

xviii "We must protest": Shenk, *The Genius in All of Us*, 29.

xviii By the 1930s, schools were testing: Ravitch, *Left Back*, 239, 368; see also Tyack, *The One Best System*, 185–90.

xviii "the limits of a child's educability": Ravitch, *Left Back*, 138; Lewis M. Terman, "The Use of Intelligence Tests in the Grading of School Children," *Journal of Educational Research* 1, no. 1 (1920): 31.

xviii Upper-track kids: Gordon Thomas Way, "Examining Testing Policy in the United States: A Comparative Historical Analysis of National Testing for Accountability Debates and Intelligence Testing Debates" (PhD diss., University of Kansas, 2014).

xviii Indeed, when the state of Maine: Michael Hurwitz, Jonathan Smith, Sunny Niu, and Jessica Howell. "The Maine Question: How Is 4-year College Enrollment Affected by Mandatory College Entrance Exams?" *Educational Evaluation and Policy Analysis* 37, no. 1 (2015): 138–59.

xviii In 2005, for instance: David Card and Laura Giuliano, "Universal Screening Increases the Representation of Low-Income and Minority Students in Gifted Education," *Proceedings of the National Academy of Sciences* 113, no. 48 (2016): 13678–83.

xix Some psychologists doubt: For example, see Howard Gardner, *Multiple Intelligences: New Horizons* (New York: Basic Books, 2006).

xix Starkly drawn test questions: Robert J. Sternberg, *Beyond IQ: A Triarchic Theory of Human Intelligence* (CUP Archive, 1985).

xix the lower a family's socioeconomic status: E. Turkenheim, A. Haley, M. Waldron, B. D'Onofrio, and I. I. Gottesman, "Socioeconomic Status Modifies Heritability of IQ in Young Children," *Psychological Science* 14, no 6 (2003): 623–28.

xix Pollution, for instance, such as lead: Aaron Reuben, Avshalom Caspi, Daniel W. Belsky, Jonathan Broadbent, Honalee Harrington, Karen Sugden, Renate M. Houts, Sandhya Ramrakha, Richie Poulton, and Terrie E. Moffitt, "Association of Childhood Blood Lead Levels with Cognitive Function and Socioeconomic Status at Age 38 Years and with IQ Change and Socioeconomic Mobility Between Childhood and Adulthood," *Journal of the American Medical Association* 317, no. 12 (2017): 1244–51.

xix childhood malnutrition: Alan Lucas, "Long-Term Programming Effects of Early Nutrition—Implications for the Preterm Infant," *Journal of Perinatology* 25, no. S2 (2005).

xix childhood abuse and neglect: Michael D. De Bellis and Abigail Zisk, "The Biological Effects of Childhood Trauma," *Child and Adolescent Psychiatric Clinics* 23, no. 2 (2014): 185–222.

xix lack of sleep: Namni Goel, Hengyi Rao, Jeffrey S. Durmer, and David F. Dinges, "Neurocognitive Consequences of Sleep Deprivation," *Seminars in Neurology* 29, no. 4 (2009): 320–39.

xix acute stress: Clancy Blair and C. Cybele Raver, "Poverty, Stress, and Brain Development: New Directions for Prevention and Intervention," *Academic Pediatrics* 16, no. 3 (2016): S30–S36; Gary W. Evans and Michelle A. Schamberg, "Childhood Poverty, Chronic Stress, and Adult Working Memory," *Proceedings of the National Academy of Sciences* 106, no. 16 (2009): 6545–49.

xix every additional year of schooling: Christopher Winship and Sanders Korenman, "Does Staying in School Make You Smarter? The Effect of Education on IQ in *The Bell Curve*," in *Intelligence, Genes, and Success* (New York: Springer, 1997), 215–34.

xix the simple act of teaching students: David S. Yeager et al., "A National Experiment Reveals Where a Growth Mindset Improves Achievement," *Nature* 573, no. 7774 (2019): 364–69.

xx "Generations of affluent people": Carl Kaestle, "Testing Policy in the United States: A Historical Perspective," the Gordon Commission, Educational Testing Service (ETS), 2013; https://www.ets.org/Media/Research/pdf/kaestle_testing_policy_us_historical_perspective.pdf.

xx In the summer of 2019: Richard D. Kahlenberg, "An Imperfect SAT

Adversity Score Is Better Than Just Ignoring Adversity," *The Atlantic*, May 25, 2019.

xx In 1993, the College Board: Mary Jordan, "SAT Changes Name, but It Won't Score 1,600 with Critics," *Washington Post*, March 27, 1993.

xx and then, in 1997: Peter Applebome, "Insisting It's Nothing, Creator Says SAT, not SAT," *New York Times*, April 2, 1997.

xx the matter of stereotype threat: Charlotte R. Pennington, Derek Heim, Andrew R. Levy, and Derek T. Larkin, "Twenty Years of Stereotype Threat Research: A Review of Psychological Mediators," *Public Library of Science One* 11, no. 1 (2016): e0146487.

xx a group of boys and girls who: Jonathan Taylor, "Fairness to Gifted Girls: Admissions to New York City's Elite Public High Schools," *Journal of Women and Minorities in Science and Engineering* 25, no 1 (2019): 75–91; Jill Barshay, "The Problem with High-Stakes Testing and Women in STEM," *Hechinger Report*, January 7, 2019.

xxi occupying girls' cognitive resources: Claude M. Steele and Joshua Aronson, "Stereotype Threat and the Test Performance of Academically Successful African Americans," *Journal of Personality and Social Psychology* 69, no 5 (1995): 797–811.

xxiii "The one thing": Jonçich, *The Sane Positivist*, 322.

xxiii In fact, the learning mechanism: For a notable dissent, see Gallistel and King, *Memory and the Computational Brain*.

xxiv a high-rise of sorts: For a notable dissent, see Varela, Rosch, and Thompson, *The Embodied Mind*.

xxvii Meanwhile, at a time: Raj Chetty, Nathaniel Hendren, Patrick Kline, Emmanuel Saez, and Nicholas Turner, "Is the United States Still a Land of Opportunity? Recent Trends in Intergenerational Mobility," *American Economic Review* 104, no. 5 (2014): 141–47.

xxvii we need new ways to bolster: Chetty et al., "Mobility Report Cards," 2017.

xxvii "These findings suggest": Alex Bell, Raj Chetty, Xavier Jaravel, Neviana Petkova, and John Van Reenen, "Who Becomes an Inventor in America? The Importance of Exposure to Innovation," *Quarterly Journal of Economics* 134, no. 2 (2018): 647–713.

xxvii Scientists, meanwhile: Benjamin Jones, E. J. Reedy, and Bruce A. Weinberg, "Age and Scientific Genius," NBER Working Paper 19866, National Bureau of Economic Research (2014).

PART ONE: LEARNING IS SCIENCE AND SCIENCE IS LEARNING

I. THE LEARNING DIVIDE

7 "America's foremost philosopher": "Dr. John Dewey Dead at 92; Philosopher a Noted Liberal," *New York Times*, June 2, 1952.

7 As passive recipients: Dewey, *The Child and the Curriculum*, 24; Labaree, *The Trouble with Ed Schools*, 131.

10 "The eggheads don't get slowed up": Ferster, *Teaching Machines*, 160.

10 less effective in the wild: Jason K. McDonald, "The Rise and Fall of Programmed Instruction: Informing Instructional Technologists Through a Study of the Past" (PhD diss., Brigham Young University, 2003), 20.

10 the public found them creepy: Sydney Katz, "Some of Johnny's Best Teachers Are Machines," *Maclean's*, March 24, 1962.

10 Once the novelty wore off: McDonald, "The Rise and Fall of Programmed Instruction," 42.

11 "I have often argued": Ellen Condliffe Lagemann, "The Plural Worlds of Educational Research," *History of Education Quarterly* 29, no. 2 (1989): 185–214.

12 Holding fast to one side: Kliebard, *The Struggle for the American Curriculum: 1893–1958*, 5; Turbayne, *The Myth of Metaphor*.

12 the otherwise egalitarian Dewey: Kliebard, *The Struggle for the American Curriculum*, 57.

12 tracing metaphorical connections: Thomas Fallace, "Recapitulation Theory and the New Education: Race, Culture, Imperialism, and Pedagogy, 1894–1916," *Curriculum Inquiry* 42, no. 4 (2012): 510–33.

12 Hall took it further still: Kliebard, *The Struggle for the American Curriculum*, 39.

13 "reason is only dawning": Ibid., 43.

13 Dewey's debut: Ibid., 47; John Dewey, "Interest in Relation to Training of the Will," *Herbart Yearbook for 1895*, 2nd supp., 1896, 209–46.

13 The timing was perfect: Lagemann, *An Elusive Science*, 42.

14 "The school is the one form": Ibid.

14 "The native and unspoiled attitude": Dewey, *How We Think*, preface.

14 "short-circuit for the individual": Ibid., 156.

14 "as a preparation": Dewey and Boydston, *The Early Works of John Dewey, 1882–1898, Volume 5: Early Essays, 1895–1898*, 224.

14 And the school as a whole: Kliebard, *The Struggle for the American Curriculum*, 55.

15 "the same relation to the work": Lagemann, *An Elusive Science*, 49; John Dewey, "Pedagogy as a University Discipline," in Dewey and Boydston, *The Early Works of John Dewey, 1882–1898, Volume 5: Early Essays, 1895–1898*, 437.

15 "a natural avenue": Kliebard, *The Struggle for the American Curriculum*, 62; John Dewey, "The University Elementary School: History and Character," *University [of Chicago] Record*, 2, 72.

15 "eternally set off": Kliebard, *The Struggle for the American Curriculum*, 64; John Dewey, "The Psychological Aspect of the School Curriculum," *Educational Review* 13 (1897): 361.

15 "with the instruments": Kliebard, *The Struggle for the American Curriculum*, 69; Dewey, *The School and Society*, 44.

15 "a larger society": Kliebard, *The Struggle for the American Curriculum*, 69; Dewey, *The School and Society*, 44.

16 his days conducting original education research: Westbrook, *John Dewey and American Democracy*, 113.

16 "I just cannot understand Dewey!": Jonçich, *The Sane Positivist*, 3.

16 When Dewey read James's articles: Lagemann, *An Elusive Science*, 62.

17 "stimulating, more so": Edward L. Thorndike, "Edward Lee Thorndike," in Murchison, *A History of Psychology in Autobiography*, vol. 3, 263–70.

17 Funnily enough: Jonçich, *The Sane Positivist*, 48.

17 Harvard told him: Stephen Tomlinson, "Edward Lee Thorndike and John Dewey on the Science of Education," *Oxford Review of Education* 23, no. 3 (1997): 365–83.

18 "nuisance to Mrs. James": Thorndike, "Edward Lee Thorndike," 263–70.

18 Before completing his degree: Lagemann, *An Elusive Science*, 57.

18 In Manhattan, he set up shop: W. Cumming, "A Review of Geraldine Jonçich's *The Sane Positivist: A Biography of Edward L. Thorndike*," *Journal of the Experimental Analysis of Behavior* 72, no. 3 (1999): 429.

18 "would have shamed": R. L. Thorndike, "Edward Thorndike: A Personal and Professional Appreciation," in Kimble et al., *Portraits of Pioneers in Psychology*, vol. 1, 139–51.

18 "You'd like to see": Jonçich, *The Sane Positivist*, 139.

19 By all rights, his report: A. Charles Catania, "Thorndike's Legacy: Learning, Selection, and the Law of Effect," *Journal of the Experimental Analysis of Behavior* 72, no. 3 (1999): 425–28.

19 "I've got some theories": Jonçich, *The Sane Positivist*, 146.

19 He found this ludicrous: Ibid., 142–43.

19 The most important of these: Gray, *Psychology*, 108–9.
20 "the main, and perhaps the only": Jonçich, *The Sane Positivist*, 352; Thorndike, *The Psychology of Learning*, 16.
20 Ultimately, it wasn't these experiments: Jamie Chamberlin, "Notes on a Scandal," *Monitor on Psychology* 43, no. 9 (2012).
20 He spent the rest of his career: "John B. Watson," *Ad Age*, 1999.
21 Skinner's pigeon-training efforts: Joseph Stromburg, "BF Skinner's Pigeon-Guided Rocket," Smithsonian.com, August 18, 2011.
21 "There is a genetic connection": Ferster, *Teaching Machines*, 69.
21 Skinner's research suggested: Burrhus Frederic Skinner, "The Technology of Teaching," BF Skinner Foundation, 2016, 35–36.
21 teaching machines and related patents: Bill Ferster, *Teaching Machines* (Baltimore: Johns Hopkins University Press, 2014), 55–60.
22 His Law of Effect provided: "Exploring the Role of Teachers College in International Education," *International Education News*, May 30, 2019.
22 *administrative progressives:* Tyack, *The One Best System*, 127.
22 urban migration: *United States Summary: 2010*, prepared by the United States Bureau of the Census (Washington, DC, September 2012), 13, 21; https://www.census.gov/prod/cen2010/cph-2-1.pdf.
22 recessions: Labaree, *Someone Has to Fail*, 88.
22 the new economic concept: Graebner, *A History of Retirement*, 15–16.
22 the concomitant rise: Labaree, *Someone Has to Fail*, 88.
23 "the tools necessary to atomise": Stephen Tomlinson, "Edward Lee Thorndike and John Dewey on the Science of Education," *Oxford Review of Education* 23, no. 3 (1997): 365–83.
23 the faculty of "judgment": Jonçich, *The Sane Positivist*, 271.
23 with particular zeal: Edward L. Thorndike and Robert S. Woodworth, "The Influence of Improvement in One Mental Function upon the Efficiency of Other Functions: III. Functions Involving Attention, Observation and Discrimination," *Psychological Review* 8, no. 6 (1901): 553.
23 "the home, the farm": *Reorganization of Science in Secondary Schools: A Report*, National Education Association of the United States (Washington, DC, 1920).
24 Thorndike, who made a fortune: Richard E. Mayer, "E. L. Thorndike's Enduring Contributions to Educational Psychology," in Zimmerman and Schunk, *Educational Psychology*, 140.
24 "the effect of every possible stimulus": Thorndike, *The Principles of Teaching*, 9.

24 As Ellen Condliffe Lagemann has pointed out: Lagemann, *An Elusive Science*, 61.

24 gender was a major component: See, for example, Goldstein, *The Teacher Wars.*

24 To this new administrative class: James M. Heffernan, "The Credibility of the Credit Hour: The History, Use, and Shortcomings of the Credit System," *Journal of Higher Education* 44, no. 1 (1973): 61–72.

25 "Grammar school, high school, and college": Thorndike, *Notes on Child Study*, 292.

25 "such a training": Kliebard, *The Struggle for the American Curriculum*, 93; E. L. Thorndike, "The Opportunity of the High Schools," *The Bookman*, October 1906, 180.

25 "Those who have the most": Edward L. Thorndike, "Mental Discipline in High School Studies," *Journal of Educational Psychology* 15, no. 2 (1924): 83.

25 "there may be some question": Kliebard, *The Struggle for the American Curriculum*, 91.

25 he did personally come up with: Tomlinson, "Edward Lee Thorndike and John Dewey on the Science of Education."

26 "procedure which under the title": Ravitch, *Left Back*, 151; John Dewey, "Individuality, Equality, and Superiority," *New Republic* 33 (December 13, 1922): 61–63.

26 "became a self-fulfilling prophecy": Ravitch, *Left Back*, 156.

27 "Galton have influenced me": Edward L. Thorndike, "Edward Lee Thorndike," in Murchison, *A History of Psychology in Autobiography*, vol. 3.

27 Terman, a particularly virulent: Ravitch, *Left Back*, 135.

27 Galton Society: Ibid., 143.

27 "One sure service": Thorndike, *Human Nature and the Social Order*, 957.

27 What's more, marginalized populations: See Fish, *Race and Intelligence*, 241–78.

27 the only plausible explanation: Ned Block, "How Heritability Misleads About Race," *Boston Review* 20, no. 6 (1996): 30–35.

27 Or, as *Vox*'s Ezra Klein has put it: Ezra Klein, "Sam Harris, Charles Murray, and the Allure of Race Science," *Vox*, March 27, 2018.

28 Thorndike used an index: Tomlinson, "Edward Lee Thorndike and John Dewey on the Science of Education"; Thorndike, *Human Nature and the Social Order*, 957.

II. LAYER ONE: SLUG CELLS AND SCHOOL BELLS

30 In one study from 2013: Jonathan A. Susser and Jennifer McCabe, "From the Lab to the Dorm Room: Metacognitive Awareness and Use of Spaced Study," *Instructional Science* 41, no. 2 (2013): 345–63.

31 When researchers broke: Nicholas J. Cepeda, Noriko Coburn, Doug Rohrer, John T. Wixted, Michael C. Mozer, and Harold Pashler, "Optimizing Distributed Practice: Theoretical Analysis and Practical Implications," *Experimental Psychology* 56, no. 4 (2009): 236.

31 Want to learn math?: Doug Rohrer and Kelli Taylor, "The Effects of Overlearning and Distributed Practise on the Retention of Mathematics Knowledge," *Applied Cognitive Psychology* 20, no. 9 (2006): 1209–24.

31 Researchers have observed similar results: Haley A. Vlach and Catherine M. Sandhofer, "Distributing Learning over Time: The Spacing Effect in Children's Acquisition and Generalization of Science Concepts," *Child Development* 83, no. 4 (2012): 1137–44.

31 Spacing has been shown to work: David A. Balota, Janet M. Duchek, and Ronda Paullin, "Age-Related Differences in the Impact of Spacing, Lag, and Retention Interval," *Psychology and Aging* 4, no. 1 (1989): 3.

31 children, and even infants: Christopher D. Smith and Damian Scarf, "Spacing Repetitions over Long Timescales: A Review and a Reconsolidation Explanation," *Frontiers in Psychology* 8 (2017): 962.

31 unplanned-for, incidental learning: Arthur M. Glenberg and Thomas S. Lehmann, "Spacing Repetitions over 1 Week," *Memory & Cognition* 8, no. 6 (1980): 528–38.

31 even for motor skills: G. Rubin-Rabson, "Studies in the Psychology of Memorizing Piano Music: II. A Comparison of Massed and Distributed Practice," *Journal of Educational Psychology* 31, no. 4 (1940): 270–84.

31 novice golfers: Teresa K. Dail and Robert W. Christina, "Distribution of Practice and Metacognition in Learning and Long-Term Retention of a Discrete Motor Task," *Research Quarterly for Exercise and Sport* 75, no. 2 (2004): 148–55.

31 In one 2006 study: Carol-Anne E. Moulton, Adam Dubrowski, Helen MacRae, Brent Graham, Ethan Grober, and Richard Reznick, "Teaching Surgical Skills: What Kind of Practice Makes Perfect?: A Randomized, Controlled Trial," *Annals of Surgery* 244, no. 3 (2006): 400.

31 Wide-ranging though: K. Matthew Lattal, "Trial and Intertrial Dura-

tions in Pavlovian Conditioning: Issues of Learning and Performance," *Journal of Experimental Psychology: Animal Behavior Processes* 25, no. 4 (1999): 433.

31 Fruit flies can be taught: J. C. P. Yin, M. Del Vecchio, H. Zhou, and T. Tully, "CREB as a Memory Modulator: Induced Expression of a dCREB2 Activator Isoform Enhances Long-Term Memory in Drosophila," *Cell* 81, no. 1 (1995): 107–15.

32 The human brain contains: Suzana Herculano-Houzel, "The Human Brain in Numbers: A Linearly Scaled-Up Primate Brain," *Frontiers in Human Neuroscience* 3 (2009): 31.

32 Repetition, he observed: Hermann Ebbinghaus, "Memory: A Contribution to Experimental Psychology," *Annals of Neurosciences* 20, no. 4 (2013): 155.

33 It was probably his student: Dehn, *Working Memory and Academic Learning*, 10.

33 "What an organism does": Diego Zilio, "Filling the Gaps: Skinner on the Role of Neuroscience in the Explanation of Behavior," *Behavior and Philosophy* 41 (2013): 33–59; Skinner, *About Behaviorism*, 219.

34 "The cerebral cortex": Edward G. Jones, "Santiago Ramon y Cajal and the Croonian Lecture, March 1894," *Trends in Neurosciences* 17, no. 5 (1994): 190–92.

35 A given cortical neuron: Aertsen and Braitenberg, *Information Processing in the Cortex*, 9.

35 "Neurons that fire together": For example, Carla J. Shatz, "The Developing Brain," *Scientific American* 267, no. 3 (1992): 60–67.

35 Behaviorism-inclined researchers: Eric R. Kandel, "The Molecular Biology of Memory Storage: A Dialogue Between Genes and Synapses," *Science* 294, no. 5544 (2001): 1030–38.

36 The longest axon in the animal kingdom: Mathew J. Wedel, "A Monument of Inefficiency: The Presumed Course of the Recurrent Laryngeal Nerve in Sauropod Dinosaurs," *Acta Palaeontologica Polonica* 57, no. 2 (2011): 251–57.

36 After a maddening early experience: Kandel, *In Search of Memory*, 107.

37 "but I found the bang!": Ibid., 107–8.

37 In short order: Ibid., 139.

37 That suggested an intriguing possibility: Ibid., 142.

37 "Few self-respecting neurophysiologists": Ibid., 143.

37 But Kandel knew: Eric R. Kandel, "Eric Kandel—Biographical," NobelPrize.org, 2000; https://www.nobelprize.org/prizes/medicine/2000/kandel/biographical.

38 "It occurred to me": Ibid.

38 Sometimes, the locomotive meets: Leonid L. Moroz, "Aplysia," *Current Biology: CB* 21, no. 2 (2011): R60.

39 Kandel achieved: Kandel, *In Search of Memory*, 169–70.

40 This principle, however: Ibid., 171.

40 The team settled: Ibid., 194.

40 Classical conditioning proved tougher: Ibid., 201.

41 "The *potential* for many": Ibid., 202.

41 When you first smelled: Caroline Bushdid, Marcelo O. Magnasco, Leslie B. Vosshall, and Andreas Keller, "Humans Can Discriminate More than 1 Trillion Olfactory Stimuli," *Science* 343, no. 6177 (2014): 1370–72.

42 get used to a smell: Michelle T. Tong, Shane T. Peace, and Thomas A. Cleland, "Properties and Mechanisms of Olfactory Learning and Memory," *Frontiers in Behavioral Neuroscience* 8 (2014): 238.

44 In 1963, a team led by: Josefa B. Flexner, Louis B. Flexner, and Eliot Stellar, "Memory in Mice as Affected by Intracerebral Puromycin," *Science* 141, no. 3575 (1963): 57–59.

44 Kandel's team replicated: James H. Schwartz, Vincent F. Castellucci, and Eric R. Kandel, "Functioning of Identified Neurons and Synapses in Abdominal Ganglion of Aplysia in Absence of Protein Synthesis," *Journal of Neurophysiology* 34, no. 6 (1971): 939–53.

44 Intuitively enough: Marcello Brunelli, V. Castellucci, and E. R. Kandel, "Synaptic Facilitation and Behavioral Sensitization in Aplysia: Possible Role of Serotonin and Cyclic AMP," *Science* 194, no. 4270 (1976): 1178–81; E. R. Kandel, M. Brunelli, J. Byrne, and V. Castellucci, "A Common Presynaptic Locus for the Synaptic Changes Underlying Short-Term Habituation and Sensitization of the Gill-Withdrawal Reflex in Aplysia," *Cold Spring Harbor Symposia on Quantitative Biology*, vol. 40 (Cold Spring Harbor Laboratory Press, 1976), 465–82.

45 Stunningly, the resulting boost: Kandel, *In Search of Memory*, 282–83.

45 Like a well-preserved memory: Eric R. Kandel, Yadin Dudai, and Mark R. Mayford, "The Molecular and Systems Biology of Memory," *Cell* 157, no. 1 (2014): 163–86.

45 Today, LTP isn't the only candidate: Ibid.

45 Even wilder, whole new spines: Costandi, *Neuroplasticity*, 63; Raphael Lamprecht and Joseph LeDoux, "Structural Plasticity and Memory," *Nature Reviews Neuroscience* 5, no. 1 (2004): 45.

45 The indirect evidence: Wickliffe C. Abraham, Barbara Logan, Jeffrey M. Greenwood, and Michael Dragunow, "Induction and Experience-Dependent Consolidation of Stable Long-Term Potentiation Lasting

Months in the Hippocampus," *Journal of Neuroscience* 22, no. 21 (2002): 9626–34.

45 its effects can be impaired: Carol A. Barnes, "Long-Term Potentiation and the Ageing Brain," *Philosophical Transactions of the Royal Society of London. Series B: Biological Sciences* 358, no. 1432 (2003): 765–72.

45 Meanwhile, when neuroscientists tailored: Benedict C. Albensi, Derek R. Oliver, Justin Toupin, and Gary Odero, "Electrical Stimulation Protocols for Hippocampal Synaptic Plasticity and Neuronal Hyper-Excitability: Are They Effective or Relevant?" *Experimental Neurology* 204, no. 1 (2007): 1–13.

46 In this model: Kandel, *In Search of Memory*, 264–65.

46 intensifies its physiological effects: Enikö A. Kramár, Alex H. Babayan, Cristin F. Gavin, Conor D. Cox, Matiar Jafari, Christine M. Gall, Gavin Rumbaugh, and Gary Lynch, "Synaptic Evidence for the Efficacy of Spaced Learning," *Proceedings of the National Academy of Sciences* 109, no. 13 (2012): 5121–26.

46 on both upstream: Bulent Ataman, James Ashley, Michael Gorczyca, Preethi Ramachandran, Wernher Fouquet, Stephan J. Sigrist, and Vivian Budnik, "Rapid Activity-Dependent Modifications in Synaptic Structure and Function Require Bidirectional Wnt signaling," *Neuron* 57, no. 5 (2008): 705–18.

46 and downstream: Gang-Yi Wu, Karl Deisseroth, and Richard W. Tsien, "Spaced Stimuli Stabilize MAPK Pathway Activation and Its Effects on Dendritic Morphology," *Nature Neuroscience* 4, no. 2 (2001): 151.

46 If information encountered repeatedly: Enikö A. Kramár, Alex H. Babayan, Cristin F. Gavin, Conor D. Cox, Matiar Jafari, Christine M. Gall, Gavin Rumbaugh, and Gary Lynch, "Synaptic Evidence for the Efficacy of Spaced Learning," *Proceedings of the National Academy of Sciences* 109, no. 13 (2012): 5121–26.

50 It's not known: Jürgen Kornmeier and Zrinka Sosic-Vasic, "Parallels Between Spacing Effects During Behavioral and Cellular Learning," *Frontiers in Human Neuroscience* 6 (2012): 203.

50 We know that: Cepeda et al., "Optimizing Distributed Practice," 2009; Nicholas J. Cepeda, Harold Pashler, Edward Vul, John T. Wixted, and Doug Rohrer, "Distributed Practice in Verbal Recall Tasks: A Review and Quantitative Synthesis," *Psychological Bulletin* 132, no. 3 (2006): 354.

51 When tested afterward: Kelli Taylor and Doug Rohrer, "The Effects of Interleaved Practice," *Applied Cognitive Psychology* 24, no. 6 (2010): 837–48.

52 The principle works: Kandel and Mack, *Principles of Neural Science*, 470–72.

53 As judicious observers have pointed out: Tomonori Takeuchi, Adrian J. Duszkiewicz, and Richard G. M. Morris, "The Synaptic Plasticity and Memory Hypothesis: Encoding, Storage and Persistence," *Philosophical Transactions of the Royal Society B* 369, no. 1633 (2014).

53 Later, laid out on: Xu Liu, Steve Ramirez, Petti T. Pang, Corey B. Puryear, Arvind Govindarajan, Karl Deisseroth, and Susumu Tonegawa, "Optogenetic Stimulation of a Hippocampal Engram Activates Fear Memory Recall," *Nature* 484, no. 7394 (2012): 381.

54 Since its invention in 2004: Edward S. Boyden, Feng Zhang, Ernst Bamberg, Georg Nagel, and Karl Deisseroth, "Millisecond-Timescale, Genetically Targeted Optical Control of Neural Activity," *Nature Neuroscience* 8, no. 9 (2005): 1263–68.

54 In a study published in 2015: Susumu Tonegawa, Michele Pignatelli, Dheeraj S. Roy, and Tomás J. Ryan, "Memory Engram Storage and Retrieval," *Current Opinion in Neurobiology* 35 (2015): 101–9.

55 a pair of papers in 2016: Dheeraj S. Roy, Autumn Arons, Teryn I. Mitchell, Michele Pignatelli, Tomás J. Ryan, and Susumu Tonegawa, "Memory Retrieval by Activating Engram Cells in Mouse Models of Early Alzheimer's Disease," *Nature* 531, no. 7595 (2016): 508–12.

55 and 2017: Takashi Kitamura, Sachie K. Ogawa, Dheeraj S. Roy, Teruhiro Okuyama, Mark D. Morrissey, Lillian M. Smith, Roger L. Redondo, and Susumu Tonegawa, "Engrams and Circuits Crucial for Systems Consolidation of a Memory," *Science* 356, no. 6333 (2017): 73–78.

55 Perhaps, as Roy and Tonegawa's team suggested: Mu-ming Poo, Michele Pignatelli, Tomás J. Ryan, Susumu Tonegawa, Tobias Bonhoeffer, Kelsey C. Martin, Andrii Rudenko, et al., "What Is Memory? The Present State of the Engram," *BMC Biology* 14, no. 1 (2016): 40.

56 Sossin, meanwhile, has gone so far: Wayne S. Sossin, "Memory Synapses Are Defined by Distinct Molecular Complexes: A Proposal," *Frontiers in Synaptic Neuroscience* 10 (2018): 5.

III. LAYER TWO: SYSTEMS WITHIN SYSTEMS

58 Each individual voxel: Assuming 7,000 synapses per cortical neuron, 630,000 neurons per 3mm3 fMRI voxel, 37 acres of maintained turf per 9-hole golf course, and 3,000 blades of grass per square foot.

59 They isolated a key synapse: Toshihide Hige, Yoshinori Aso, Meh-

rab N. Modi, Gerald M. Rubin, and Glenn C. Turner, "Heterosynaptic Plasticity Underlies Aversive Olfactory Learning in Drosophila," *Neuron* 88, no. 5 (2015): 985–98.

61 "I sometimes feel": Karl S. Lashley, "In Search of the Engram," *Symposiums of the Society of Experimental Biology*, no. 4 (1950): 454–82.

62 Instead, researchers hypothesized: Larry R. Squire and John T. Wixted, "The Cognitive Neuroscience of Human Memory Since HM," *Annual Review of Neuroscience* 34 (2011): 259–88.

62 Of the non-hippocampal regions: Larry R. Squire and John T. Wixted, "The Cognitive Neuroscience of Human Memory Since HM," *Annual Review of Neuroscience* 34 (2011): 259–88.

63 "Huh, this was easier": Benedict Carey, "H. M., an Unforgettable Amnesiac, Dies at 82," *New York Times*, December 4, 2008.

63 the forty-seven-year-old woman: Daniel L. Schacter, "Implicit Memory: History and Current Status," *Journal of Experimental Psychology: Learning, Memory, and Cognition* 13, no. 3 (1987): 501.

64 too much stock in priming: Wojciech Świątkowski and Benoît Dompnier, "Replicability Crisis in Social Psychology: Looking at the Past to Find New Pathways for the Future," *International Review of Social Psychology* 30, no. 1 (2017).

64 Molaison could be primed: John D. E. Gabrieli, William Milberg, Margaret M. Keane, and Suzanne Corkin, "Intact Priming of Patterns Despite Impaired Memory," *Neuropsychologia* 28, no. 5 (1990): 417–27.

64 newer, more accurate tools: John D. E. Gabrieli, Margaret M. Keane, Ben Z. Stanger, Margaret M. Kjelgaard, Suzanne Corkin, and John H. Growdon, "Dissociations Among Structural-Perceptual, Lexical-Semantic, and Event-Fact Memory Systems in Alzheimer, Amnesic, and Normal Subjects," *Cortex* 30, no. 1 (1994): 75–103.

65 one of the first fMRI images: J. W. Belliveau, D. N. Kennedy, R. C. McKinstry, B. R. Buchbinder, R. M. Weisskoff, M. S. Cohen, J. M. Vevea, T. J. Brady, and B. R. Rosen, "Functional Mapping of the Human Visual Cortex by Magnetic Resonance Imaging," *Science* 254, no. 5032 (1991): 716–19.

66 "But I needed to stop": Nancy Kanwisher, "The Quest for the FFA and Where It Led," *Journal of Neuroscience* 37, no. 5 (2017): 1056–61.

67 In a single, high-stakes take: Nancy Kanwisher, "The Neuroanatomy Lesson," April 14, 2015. YouTube video, 1:28; https://www.youtube.com/watch?v=PcbSQxJ7UrU.

68 there appears to exist: Kanwisher, "The Quest for the FFA."

69 going all the way back to 1861: Paul Broca, "Remarques sur le siège

de la faculté du langage articulé, suivies d'une observation d'aphémie (perte de la parole)," *Bulletins et Mémoires de la Société anatomique de Paris* 6 (1861): 330–57.

69 an fMRI researcher could present: Cathy J. Price, "A Review and Synthesis of the First 20 Years of PET and fMRI Studies of Heard Speech, Spoken Language and Reading," *Neuroimage* 62, no. 2 (2012): 816–47.

70 letters-in-the-abstract: Dehaene, *Reading in the Brain*, 91.

70 including *graphemes:* A fascinating new research thread suggests that graphemes in particular may be parsed not in the visual word form area (the brain's letterbox) itself, but in a nearby region dubbed the grapheme-related area. Florence Bouhali, Zoé Bézagu, Stanislas Dehaene, and Laurent Cohen, "A Mesial-to-Lateral Dissociation for Orthographic Processing in the Visual Cortex," *Proceedings of the National Academy of Sciences* 116, no. 43 (2019): 21936–46.

71 A lesion in the fusiform face area: Jason J. S. Barton, Daniel Z. Press, Julian P. Keenan, and Margaret O'Connor, "Lesions of the Fusiform Face Area Impair Perception of Facial Configuration in Prosopagnosia," *Neurology* 58, no. 1 (2002): 71–78.

71 And damage to the letterbox: Peter E. Turkeltaub, Ethan M. Goldberg, Whitney A. Postman-Caucheteux, Merisa Palovcak, Colin Quinn, Charles Cantor, and H. Branch Coslett, "Alexia Due to Ischemic Stroke of the Visual Word Form Area," *Neurocase* 20, no. 2 (2014): 230–35.

71 a very specific type of memory: Stanislas Dehaene and Laurent Cohen, "The Unique Role of the Visual Word Form Area in Reading," *Trends in Cognitive Sciences* 15, no. 6 (2011): 254–62.

71 The 5 to 12 percent: Elizabeth S. Norton, Sara D. Beach, and John D. E. Gabrieli, "Neurobiology of Dyslexia," *Current Opinion in Neurobiology* 30 (2015): 73–78.

71 multiple causes of dyslexia: John D. E. Gabrieli, "Dyslexia: A New Synergy Between Education and Cognitive Neuroscience," *Science* 325, no. 5938 (2009): 280–83.

72 Someone with dyslexia might: Norton et al., "Neurobiology of Dyslexia."

73 A super-long-distance white-matter pathway: Torkel Klingberg, Maj Hedehus, Elise Temple, Talya Salz, John D. E. Gabrieli, Michael E. Moseley, and Russell A. Poldrack, "Microstructure of Temporo-Parietal White Matter as a Basis for Reading Ability: Evidence from Diffusion Tensor Magnetic Resonance Imaging," *Neuron* 25, no. 2 (2000): 493–500.

73 left arcuate fasciculus: Bart Boets, Hans P. Op de Beeck, Maaike Vandermosten, Sophie K. Scott, Céline R. Gillebert, Dante Mantini, Jessica Bulthé, Stefan Sunaert, Jan Wouters, and Pol Ghesquière, "Intact but Less Accessible Phonetic Representations in Adults with Dyslexia," *Science* 342, no. 6163 (2013): 1251–54; Joanna A. Christodoulou, Jack Murtagh, Abigail Cyr, Tyler K. Perrachione, Patricia Chang, Kelly Halverson, Pamela Hook, Anastasia Yendiki, Satrajit Ghosh, and John D. E. Gabrieli, "Relation of White-Matter Microstructure to Reading Ability and Disability in Beginning Readers," *Neuropsychology* 31, no. 5 (2017): 508.

73 *phonological awareness:* Jason D. Yeatman, Robert F. Dougherty, Elena Rykhlevskaia, Anthony J. Sherbondy, Gayle K. Deutsch, Brian A. Wandell, and Michal Ben-Shachar, "Anatomical Properties of the Arcuate Fasciculus Predict Phonological and Reading Skills in Children," *Journal of Cognitive Neuroscience* 23, no. 11 (2011): 3304–17.

73 "There is no other human behavior": Tyler K. Perrachione, Stephanie N. Del Tufo, Rebecca Winter, Jack Murtagh, Abigail Cyr, Patricia Chang, Kelly Halverson, Satrajit S. Ghosh, Joanna A. Christodoulou, and John D. E. Gabrieli, "Dysfunction of Rapid Neural Adaptation in Dyslexia," *Neuron* 92, no. 6 (2016): 1383–97.

74 a growing contingent of researchers: Florence Bouhali, Michel Thiebaut de Schotten, Philippe Pinel, Cyril Poupon, Jean-François Mangin, Stanislas Dehaene, and Laurent Cohen, "Anatomical Connections of the Visual Word Form Area," *Journal of Neuroscience* 34, no. 46 (2014): 15402–14.

74 "the functional fate": Zeynep M. Saygin, David E. Osher, Elizabeth S. Norton, Deanna A. Youssoufian, Sara D. Beach, Jenelle Feather, Nadine Gaab, John D. E. Gabrieli, and Nancy Kanwisher, "Connectivity Precedes Function in the Development of the Visual Word Form Area," *Nature Neuroscience* 19, no. 9 (2016): 1250.

75 the surprisingly short list: Mark A. Changizi, Qiong Zhang, Hao Ye, and Shinsuke Shimojo, "The Structures of Letters and Symbols Throughout Human History Are Selected to Match Those Found in Objects in Natural Scenes," *The American Naturalist* 167, no. 5 (2006): E117–E139.

75 "Reading itself": Dehaene, *Reading in the Brain*, 303.

75 The trick: Nadine Gaab, "Identifying Risk Instead of Failure," Blog on Learning and Development, April 3, 2019; https://bold.expert/identifying-risk-instead-of-failure/.

75 it's become possible to identify: Ola Ozernov-Palchik and Nadine Gaab, "Tackling the 'Dyslexia Paradox': Reading Brain and Behavior

for Early Markers of Developmental Dyslexia," *Wiley Interdisciplinary Reviews: Cognitive Science* 7, no. 2 (2016): 156–76.

76 "a brave but often fruitless endeavor": Dehaene, *Reading in the Brain*, 247.

77 "Law of Readiness": Thorndike, *Educational Psychology*, 1–2.

77 The approach paid off: Julie J. Yoo, Oliver Hinds, Noa Ofen, Todd W. Thompson, Susan Whitfield-Gabrieli, Christina Triantafyllou, and John D. E. Gabrieli, "When the Brain Is Prepared to Learn: Enhancing Human Learning Using Real-Time fMRI," *Neuroimage* 59, no. 1 (2012): 846–52.

80 In 1994, the Carnegie Mellon psychologist: George Loewenstein, "The Psychology of Curiosity: A Review and Reinterpretation," *Psychological Bulletin* 116, no. 1 (1994): 75.

81 A small monetary reward: For example, R. Alison Adcock, Arul Thangavel, Susan Whitfield-Gabrieli, Brian Knutson, and John D. E. Gabrieli, "Reward-Motivated Learning: Mesolimbic Activation Precedes Memory Formation," *Neuron* 50, no. 3 (2006): 507–17; Bianca C. Wittmann, Björn H. Schott, Sebastian Guderian, Julietta U. Frey, Hans-Jochen Heinze, and Emrah Düzel, "Reward-Related FMRI Activation of Dopaminergic Midbrain Is Associated with Enhanced Hippocampus-Dependent Long-Term Memory Formation," *Neuron* 45, no. 3 (2005): 459–67.

81 "via dopaminergic facilitation": Matthias J. Gruber, Bernard D. Gelman, and Charan Ranganath, "States of Curiosity Modulate Hippocampus-Dependent Learning via the Dopaminergic Circuit," *Neuron* 84, no. 2 (2014): 486–96.

82 The hippocampus, in short: Ibid.; Min Jeong Kang, Ming Hsu, Ian M. Krajbich, George Loewenstein, Samuel M. McClure, Joseph Tao-yi Wang, and Colin F. Camerer, "The Wick in the Candle of Learning: Epistemic Curiosity Activates Reward Circuitry and Enhances Memory," *Psychological Science* 20, no. 8 (2009): 963–73; John E. Lisman, and Anthony A. Grace, "The Hippocampal-VTA Loop: Controlling the Entry of Information into Long-Term Memory," *Neuron* 46, no. 5 (2005): 703–13.

83 Lev Vygotsky called: Vygotsky and Cole, *Mind in Society*, 86.

84 the bothersome feeling: John D. Eastwood, Alexandra Frischen, Mark J. Fenske, and Daniel Smilek, "The Unengaged Mind: Defining Boredom in Terms of Attention," *Perspectives on Psychological Science* 7, no. 5 (2012): 482–95.

85 This approach can easily misfire: "Shakespeare Was, Like, the Ultimate Rapper," *The Onion*, August 24, 2005.

IV. LAYER THREE: REVOLUTION

92 "and ninety percent will say": Sanjoy Mahajan, "Observations on Teaching First-Year Physics," *arXiv preprint physics/0512158* (2005).

92 The question of how to coax: For example, Gerald Nelms and Ronda Leathers Dively, "Perceived Roadblocks to Transferring Knowledge from First-Year Composition to Writing-Intensive Major Courses: A Pilot Study," *WPA: Writing Program Administration* 31, no. 1–2 (2007): 214–40.

93 In American teacher-ed schools: Labaree, *Trouble with Ed Schools*, 132.

93 Depending on which of these facets: Ibid., 130.

97 He strayed beyond: C. J. Brainerd, "Jean Piaget, Learning Research, and American Education," in Zimmerman and Schunk, *Educational Psychology*, 256.

97 names like "sensorimotor": Ibid., 257.

98 Prior to the late 1950s: Ibid., 254.

99 At Harvard, these included: J. R. Hopkins, "Brown, Roger William," in Kazdin, *Encyclopedia of Psychology. Aborti–System*, vol. 1.

99 the cognitive linguist Noam Chomsky: George A. Miller, "The Cognitive Revolution: A Historical Perspective," *Trends in Cognitive Sciences* 7, no. 3 (2003): 141–44.

99 scathing 1959 review: E.g., Newmeyer, *The Politics of Linguistics*.

100 "Some think of using": Seymour Papert and Cynthia Solomon, "Twenty Things to Do with a Computer. Artificial Intelligence Memo Number 248" (1971); https://dspace.mit.edu/bitstream/handle/1721.1/5836/AIM-248.pdf?sequence=2.

101 "The construction that takes place": Papert, *The Children's Machine*, 142.

102 Papert intended the language: Mitchel Resnick et al., "Scratch: Programming for All," *Communications of the ACM* 52, no. 11 (2009): 60–67.

102 "Little by little": Papert, *Children's Machine*, 38.

103 Scratch is within striking distance: Mitchel Resnick, "Mitch Resnick: The Next Generation of Scratch Teaches More Than Coding," *EdSurge*, January 3, 2019.

105 By the late 1980s: Ravitch, *Left Back*, 446.

105 California, which had gone all in: Ibid., 447.

106 according to Sweller's study: John Sweller and Graham A. Cooper, "The Use of Worked Examples as a Substitute for Problem Solving in Learning Algebra," *Cognition and Instruction* 2, no. 1 (1985): 59–89.

106 math-related fields: Hitendra K. Pillay, "Cognitive Load and Mental

Rotation: Structuring Orthographic Projection for Learning and Problem Solving," *Instructional Science* 22, no. 2 (1994): 91–113.

106 including statistics: Fred G. Paas, "Training Strategies for Attaining Transfer of Problem-Solving Skill in Statistics: A Cognitive-Load Approach," *Journal of Educational Psychology* 84, no. 4 (1992): 429.

106 geometry: Fred G. W. C. Paas, and Jeroen J. G. Van Merriënboer, "Variability of Worked Examples and Transfer of Geometrical Problem-Solving Skills: A Cognitive-Load Approach," *Journal of Educational Psychology* 86, no. 1 (1994): 122.

106 and computer programming: John Gregory Trafton and Brian J. Reiser, "The Contributions of Studying Examples and Solving Problems to Skill Acquisition" (PhD diss., Princeton University, 1994).

107 Sweller fired off an article: John Sweller, "Cognitive Load During Problem Solving: Effects on Learning," *Cognitive Science* 12, no. 2 (1988): 257–85.

107 Back in the 1950s, Harvard's George Miller: For example, Nelson Cowan, "Metatheory of Storage Capacity Limits," *Behavioral and Brain Sciences* 24, no. 1 (2001): 154–76; Nelson Cowan, "The Magical Mystery Four: How Is Working Memory Capacity Limited, and Why?" *Current Directions in Psychological Science* 19, no. 1 (2010): 51–57; Gordon Parker, "Acta Is a Four-Letter Word," *Acta Psychiatrica Scandinavica* 126, no. 6 (2012): 476–78.

108 "refresh rate" limitation: Earl K. Miller, Mikael Lundqvist, and André M. Bastos, "Working Memory 2.0," *Neuron* 100, no. 2 (2018): 463–75.

109 Using such strategies: Nelson Cowan, "The Magical Mystery Four: How Is Working Memory Capacity Limited, and Why?" *Current Directions in Psychological Science* 19, no. 1 (2010): 51–57.

109 One upshot is: Henry Silver, Pablo Feldman, Warren Bilker, and Ruben C. Gur, "Working Memory Deficit as a Core Neuropsychological Dysfunction in Schizophrenia," *American Journal of Psychiatry* 160, no. 10 (2003): 1809–16.

109 such factors as gender: Heather Miller and Jacqueline Bichsel, "Anxiety, Working Memory, Gender, and Math Performance," *Personality and Individual Differences* 37, no. 3 (2004): 591–606.

109 race: Marleen Stelter and Juliane Degner, "Investigating the Other-Race Effect in Working Memory," *British Journal of Psychology* 109, no. 4 (2018): 777–98.

109 and socioeconomic status: Julia A. Leonard, Allyson P. Mackey, Amy S. Finn, and John D. E. Gabrieli, "Differential Effects of Socioeconomic Status on Working and Procedural Memory Systems," *Frontiers in Human Neuroscience* 9 (2015): 554.

111 Curiously enough, these come: Robert Schmidt, Maria Herrojo Ruiz, Bjorg Kilavik, Mikael Lundqvist, Philip Starr, and Adam R. Aron, "Beta Oscillations in Working Memory, Executive Control of Movement and Thought, and Sensorimotor Function," *Journal of Neuroscience* 39, no. 42 (2019): 8231–38.

112 "The basic idea is that children": Alison Gopnik, "The Theory Theory as an Alternative to the Innateness Hypothesis," *Chomsky and His Critics* (2003): 238–54.

113 As Schulz and her collaborators: Elizabeth Baraff Bonawitz, Tessa J. P. van Schijndel, Daniel Friel, and Laura Schulz, "Children Balance Theories and Evidence in Exploration, Explanation, and Learning," *Cognitive Psychology* 64, no. 4 (2012): 215–34.

113 Babies, it turns out: Alison Gopnik and Laura Schulz, "Mechanisms of Theory Formation in Young Children," *Trends in Cognitive Sciences* 8, no. 8 (2004): 371–77; Laura E. Schulz, Alison Gopnik, and Clark Glymour, "Preschool Children Learn About Causal Structure from Conditional Interventions," *Developmental Science* 10, no. 3 (2007): 322–32.

113 In one study: Hyowon Gweon and Laura Schulz, "16-Month-Olds Rationally Infer Causes of Failed Actions," *Science* 332, no. 6037 (2011): 1524; Laura Schulz, "The Surprisingly Logical Minds of Babies," TED video, 20:19; https://www.ted.com/talks/laura_schulz_the_surprisingly_logical_minds_of_babies.

V. LAYER FOUR: THINKING ABOUT THINKING

116 the Harvard psychologist Howard Gardner: Howard Gardner, "Multiple Intelligences: The First Thirty Years," *Harvard Graduate School of Education* (2011).

117 As Gardner has put it: Howard Gardner and Seana Moran, "The Science of Multiple Intelligences Theory: A Response to Lynn Waterhouse," *Educational Psychologist* 41, no. 4 (2006): 227–32.

117 To Gardner's dismay: Valerie Strauss, "Howard Gardner: 'Multiple Intelligences' Are Not 'Learning Styles,'" *Washington Post*, October 16, 2013.

117 Every large-scale review: National Research Council, *How People Learn II: Learners, Contexts, and Cultures* (National Academies Press, 2018), 137; Myron H. Dembo and Keith Howard, "Advice About the Use of Learning Styles: A Major Myth in Education," *Journal of College Reading and Learning* 37, no. 2 (2007): 101–9; Harold Pashler,

Mark McDaniel, Doug Rohrer, and Robert Bjork, "Learning Styles: Concepts and Evidence," *Psychological Science in the Public Interest* 9, no. 3 (2008): 105–19.

117 This practice revealed: "The Left Brain/Right Brain Myth," OECD Centre for Educational Research and Innovation; http://www.oecd.org/education/ceri/neuromyth6.htm.

117 but in the years since: Paul A. Howard-Jones, "Neuroscience and Education: Myths and Messages," *Nature Reviews Neuroscience* 15, no. 12 (2014): 817–24.

118 back to William James: James, *The Energies of Men*, 323. James argued that "we are making use of only a small part of our possible mental and physical resources."

118 This process molds the brain: John T. Bruer, "Education and the Brain: A Bridge Too Far," *Educational Researcher* 26, no. 8 (1997): 4–16.

118 when it delivers the finishing touches: Lynn D. Selemon, "A Role for Synaptic Plasticity in the Adolescent Development of Executive Function," *Translational Psychiatry* 3, no. 3 (2013): e238.

118 "With the right input": Sharon Begley, "Your Child's Brain," *Newsweek*, February 19, 1996, 54–57.

118 But in the mid-1990s: Elena Pasquinelli, "Neuromyths: Why Do They Exist and Persist?," *Mind, Brain, and Education* 6, no. 2 (2012): 89–96.

120 The simple task became: John A. McGeoch, "Forgetting and the Law of Disuse," *Psychological Review* 39, no. 4 (1932): 352.

122 What was more surprising: Robert A. Bjork, "Learning and Short-Term Retention of Paired Associates in Relation to Specific Sequences of Interpresentation Intervals" (PhD diss., Stanford University, 1966).

123 "if we remembered everything": James, *Psychology*, 300.

123 Without forgetting, Robert predicted: Robert A. Bjork, "On the Symbiosis of Remembering, Forgetting, and Learning," in Benjamin, ed., *Successful Remembering and Successful Forgetting*, 3; Robert A. Bjork, "Theoretical Implications of Directed Forgetting," in *Coding Processes in Human Memory*, ed. Arthur W. Melton and Edwin Martin (New York: V. H. Winston, 1972), 218.

123 The first of these involved: Robert A. Bjork and Ted W. Allen, "The Spacing Effect: Consolidation or Differential Encoding?," *Journal of Verbal Learning and Verbal Behavior* 9, no. 5 (1970): 567–72.

123 Robert also tried physically moving: Steven M. Smith, Arthur Glenberg, and Robert A. Bjork, "Environmental Context and Human Memory," *Memory & Cognition* 6, no. 4 (1978): 342–53.

123 And he explored spacing out: T. Landauer and R. A. Bjork, "Optimum

Rehearsal Patterns and Name Learning," in *Practical Aspects of Memory*, ed. M. M. Gruneberg, P. E. Morris, and R. N. Sykes (Cambridge, MA: Academic Press, 1978), 625–32.

123 compared to a control group: Robert A. Bjork, "Theoretical Implications of Directed Forgetting," in *Coding Processes in Human Memory*, ed. Arthur W. Melton and Edwin Martin (New York: V. H. Winston, 1972), 217–35.

125 The *act* of retrieving items: Robert A. Bjork and Ralph E. Geiselman, "Constituent Processes in the Differentiation of Items in Memory," *Journal of Experimental Psychology: Human Learning and Memory* 4, no. 4 (1978): 347.

129 In a classic study from 1978: Robert Kerr and Bernard Booth, "Specific and Varied Practice of Motor Skill," *Perceptual and Motor Skills* 46, no. 2 (1978): 395–401.

130 a growing list of studies: E.g., John B. Shea and Robyn L. Morgan, "Contextual Interference Effects on the Acquisition, Retention, and Transfer of a Motor Skill," *Journal of Experimental Psychology: Human Learning and Memory* 5, no. 2 (1979): 179; Timothy D. Lee and Richard A. Magill, "The Locus of Contextual Interference in Motor-Skill Acquisition," *Journal of Experimental Psychology: Learning, Memory, and Cognition* 9, no. 4 (1983): 730.

130 The former made faster progress: Sinah Goode and Richard A. Magill, "Contextual Interference Effects in Learning Three Badminton Serves," *Research Quarterly for Exercise and Sport* 57, no. 4 (1986): 308–14.

130 One motor-learning idea: Timothy D. Lee and Richard A. Magill, "The Locus of Contextual Interference in Motor-Skill Acquisition," *Journal of Experimental Psychology: Learning, Memory, and Cognition* 9, no. 4 (1983): 730; Robert A. Bjork, "On the Symbiosis of Remembering, Forgetting, and Learning," in Benjamin, ed., *Successful Remembering and Successful Forgetting*, 11.

130 The effect is more effortful retrieval: Bjork, "On the Symbiosis of Remembering, Forgetting, and Learning," 12.

131 research into pretesting: E.g., Richard C. Anderson and W. Barry Biddle, "On Asking People Questions About What They Are Reading," in *Psychology of Learning and Motivation*, vol. 9 (Cambridge, MA: Academic Press, 1975), 89–132; Janet H. Kane and Richard C. Anderson, "Depth of Processing and Interference Effects in the Learning and Remembering of Sentences," *Journal of Educational Psychology* 70, no. 4 (1978): 626; Michael Pressley, Robbi Tanenbaum, Mark A. McDaniel, and Eileen Wood, "What Happens When University Students Try to

Answer Prequestions That Accompany Textbook Material?," *Contemporary Educational Psychology* 15, no. 1 (1990): 27–35.

131 For instance, answering a question: Jeri L. Little and Elizabeth Ligon Bjork, "Multiple-Choice Pretesting Potentiates Learning of Related Information," *Memory & Cognition* 44, no. 7 (2016): 1085–1101.

132 "no evidence of the existence": W. E. Leary, "Army's Learning Panel Urges Offbeat Studies," *New York Times*, December 4, 1987.

132 The idea of metacognition: National Research Council, *Learning, Remembering, Believing: Enhancing Human Performance* (Washington, DC: National Academies Press, 1994).

132 Although both Thorndike: Thorndike, *The Psychology of Learning*.

132 and Dewey: Dewey, *How We Think*.

132 In the first modern metacognition study: Joseph T. Hart, "Memory and the Feeling-of-Knowing Experience," *Journal of Educational Psychology* 56, no. 4 (1965): 208.

133 The follow-up research: Robert A. Bjork, John Dunlosky, and Nate Kornell, "Self-Regulated Learning: Beliefs, Techniques, and Illusions," *Annual Review of Psychology* 64 (2013): 417–44.

133 This speedometer can: Robert Ariel, John Dunlosky, and Heather Bailey, "Agenda-Based Regulation of Study-Time Allocation: When Agendas Override Item-Based Monitoring," *Journal of Experimental Psychology: General* 138, no. 3 (2009); Nate Kornell and Janet Metcalfe, "Study Efficacy and the Region of Proximal Learning Framework," *Journal of Experimental Psychology: Learning, Memory, and Cognition* 32, no. 3 (2006): 609; Keith W. Thiede and John Dunlosky, "Toward a General Model of Self-Regulated Study: An Analysis of Selection of Items for Study and Self-Paced Study Time," *Journal of Experimental Psychology: Learning, Memory, and Cognition* 25, no. 4 (1999): 1024.

133 *hindsight bias:* Baruch Fischhoff, "Hindsight Is Not Equal to Foresight: The Effect of Outcome Knowledge on Judgment Under Uncertainty," *Journal of Experimental Psychology: Human Perception and Performance* 1, no. 3 (1975): 288.

133 *Foresight bias:* Asher Koriat and Robert A. Bjork, "Illusions of Competence in Monitoring One's Knowledge During Study," *Journal of Experimental Psychology: Learning, Memory, and Cognition* 31, no. 2 (2005): 187.

133 *stability bias:* Nate Kornell, "Failing to Predict Future Changes in Memory: A Stability Bias Yields Long-Term Overconfidence," in Benjamin, ed., *Successful Remembering and Successful Forgetting*, 365; Nate Kornell, "A Stability Bias in Human Memory," *Encyclopedia of the Sciences of Learning* (2012): 4–7.

133 If you can call a fact to mind: John S. Shaw III, "Increases in Eyewitness Confidence Resulting from Postevent Questioning," *Journal of Experimental Psychology: Applied* 2, no. 2 (1996): 126.

134 In a remarkable series of studies: Nate Kornell, Matthew G. Rhodes, Alan D. Castel, and Sarah K. Tauber, "The Ease-of-Processing Heuristic and the Stability Bias: Dissociating Memory, Memory Beliefs, and Memory Judgments," *Psychological Science* 22, no. 6 (2011): 787–94; Matthew G. Rhodes and Alan D. Castel, "Memory Predictions Are Influenced by Perceptual Information: Evidence for Metacognitive Illusions," *Journal of Experimental Psychology: General* 137, no. 4 (2008): 615.

134 "desirable difficulties": Robert A. Bjork, "Memory and Metamemory Considerations in the Training of Human Beings," in Metcalfe and Shimamura, *Metacognition*, 185–205; E. L. Bjork and R. A. Bjork, "Making Things Hard on Yourself, but in a Good Way: Creating Desirable Difficulties to Enhance Learning," in Gernsbacher, *Psychology and the Real World*, 56–64.

136 The school's "ultimate" pass rate: Greg Miller, "FIU Law Surpasses 95 Percent Bar Passage Milestone, Leads Florida on July 2019 Exam," *FIU News*, September 16, 2019.

139 in a 2013 article on metacognition: Robert A. Bjork, John Dunlosky, and Nate Kornell, "Self-Regulated Learning: Beliefs, Techniques, and Illusions," *Annual Review of Psychology* 64 (2013): 417–44.

PART TWO: MIND AND HAND

VI. VOYAGES

145 At a conference of physics educators: John Belcher, "Trends in Science Education," *MIT Faculty Newsletter* 9, no. 1 (September 1996).

145 No, they needed to be: Priscilla W. Laws, Pamela J. Rosborough, and Frances J. Poodry, "Women's Responses to an Activity-Based Introductory Physics Program," *American Journal of Physics* 67, no. S1 (1999): S32–S37.

145 "You're a student": Craig Lambert, "Twilight of the Lecture," *Harvard Magazine*, March–April 2012.

146 experiencing a physics concept: S. Beilock and S. Fischer, "From Cognitive Science to Physics Education and Back," in *Physics Education Research Conference 2013* (2013): 15–18.

148 In its coverage of the issue: Lauren E. LeBon, "Students Petition Against TEAL," *The Tech*, March 21, 2003.

149 By the end of the semester: Yehudit Judy Dori, Erin Hult, Lori Breslow, and John W. Belcher, "How Much Have They Retained? Making Unseen Concepts Seen in a Freshman Electromagnetism Course at MIT," *Journal of Science Education and Technology* 16, no. 4 (2007): 299–323.

149 To the shame: Robert H. Tai and Philip M. Sadler, "Gender Differences in Introductory Undergraduate Physics Performance: University Physics Versus College Physics in the USA," *International Journal of Science Education* 23, no. 10 (2001): 1017–37.

149 The findings tracked closely: Mercedes Lorenzo, Catherine H. Crouch, and Eric Mazur, "Reducing the Gender Gap in the Physics Classroom," *American Journal of Physics* 74, no. 2 (2006): 118–22.

150 In Taiwan, for instance: Ruey S. Shieh, "The Impact of Technology-Enabled Active Learning (TEAL) Implementation on Student Learning and Teachers' Teaching in a High School Context," *Computers & Education* 59, no. 2 (2012): 206–14.

150 At the University of Kentucky: "TEAL Classroom Reinvents Introductory Physics at UK," University of Kentucky, January 20, 2017; https://pa.as.uky.edu/teal-classroom-reinvents-introductory-physics-uk.

150 Without the right instructor: Ruey S. Shieh, Wheijen Chang, and Eric Zhi-Feng Liu, "Technology Enabled Active Learning (TEAL) in Introductory Physics: Impact on Genders and Achievement Levels," *Australasian Journal of Educational Technology* 27, no. 7 (2011).

151 In one 2016 survey: Dian Schaffhauser and Rhea Kelly, "55 Percent of Faculty Are Flipping the Classroom," *Campus Technology*, October 12, 2016.

153 Many of those soldiers: Tschurenev, *Empire, Civil Society, and the Beginnings of Colonial Education in India*, 31–62.

154 Bell demonstrated for his adoring onlookers: Jane Blackie, "Bell, Andrew (1753–1832), Church of England Clergyman and Educationist," *Oxford Dictionary of National Biography*, September 23, 2004.

155 After which the heat of the sun: Southey, Southey, and Southey, *The Life of the Rev. Andrew Bell*, 130.

155 Upon arrival in Madras: Blackie, "Bell, Andrew," 2004.

155 Its main funding sources: Meiklejohn, *An Old Educational Reformer, Dr Andrew Bell*, 24.

155 local "orphan" population: Southey, Southey, and Southey, *The Life of the Rev. Andrew Bell*, 135.

155 The school's original uniform: Ibid., 157.

156 "proved less fatal": Ibid., 160.

156 Bell assumed authority: Ibid., 172.
156 "Eureka!": Ibid., 173.
156 What he'd stumbled across: Tschurenev, *Empire, Civil Society, and the Beginnings of Colonial Education in India*, 31–62.
157 "It engages and amuses": Bell, *An Experiment in Education, Made at the Male Asylum at Egmore, near Madras: Suggesting a System by Which a School or Family May Teach Itself under the Superintendence of the Master or Parent*, 26.
157 He began paying: Southey, Southey, and Southey, *The Life of the Rev. Andrew Bell*, 174.
158 Within each class: Gladman, *School Work*, 14.
158 Teaching a concept: Bell, *An Experiment in Education*, 8–12.
159 Corporal punishment: Blackie, "Bell, Andrew (1753–1832)."
159 "Such interference prevented": Southey, Southey, and Southey, *The Life of the Rev. Andrew Bell*, 175.
159 Expenses, on a per-student basis: Bell, *The Madras School: Or, Elements of Tuition: Comprising the Analysis of an Experiment in Education, Made at the Male Asylum, Madras; with Its Facts, Proofs, and Illustrations; to Which Are Added, Extracts of Sermons Preached at Lambeth; a Sketch of a National Institution for Training Up the Children of the Poor; and a Specimen of the Mode of Religious Instruction at the Royal Military Asylum, Chelsea*, 181. The initial cost per student, Bell claimed, was 2.8 pagodas (or 10.1 rupees, assuming an exchange rate of 360 rupees to 100 pagodas). The subsequent cost per student was 1.75 pagodas (or 6.3 rupees), a savings of 38 percent.
159 "On no occasion": Bell, *The Madras School*, 185–86.
159 "We have already more than": Southey, Southey, and Southey, *The Life of the Rev. Andrew Bell*, 177.
160 "at the not immature age of 47": Meiklejohn, *An Old Educational Reformer*, 33.
160 "You may mark me": Ibid., 30.
160 Enthusiasm for the system: Ibid., 106.
160 Bell's first major biographical treatment: Southey, *The Origin, Nature, and Object, of the New System of Education*. This 1812 pamphlet was first published as an 1811 essay: Robert Southey, "Bell and Lancaster's Systems of Education," *Quarterly Review* 6, no. 11 (1811). See also Tom Duggett, "Southey's 'New System': The Monitorial Controversy and the Making of the 'Entire Man of Letters,'" *Romanticism and Victorianism on the Net* 61 (2012).
161 "those entrusted with power": Labaree, *Someone Has to Fail*, 50.
162 This push predated: Adam Laats, "Teacher Pay, Presidential Politics,

and New York's Modest Proposal of 1818," *History News Network*, April 14, 2019; https://historynewsnetwork.org/article/171717.

162 "When I perceive": Kaestle, *Joseph Lancaster and the Monitorial School Movement*, 158–59.

162 "There is as much waywardness": Ibid., 178.

163 A major reason: Edward Wall, "Joseph Lancaster and the Origins of the British and Foreign School Society" (PhD diss., Columbia University, 1966), 182–87.

163 They soon appeared throughout: Kaestle, *Joseph Lancaster*, 33.

163 Demand for public education: Ibid., 33–34.

163 he set off for Caracas: Lancaster, *Epitome of Some of the Chief Events and Transactions in the Life of Joseph Lancaster, Containing an Account of the Rise and Progress of the Lancasterian System of Education; . . . Written by Himself*, 36.

164 "Lancaster was always planning": Kaestle, *Joseph Lancaster*, 41–42.

164 In 1838, Lancaster was trampled to death: Ibid., 43.

164 "The sole merit": "Common Schools," *New-England Magazine* 3 (September 1832): 195–98.

165 "The masters of our primary schools": Cousin and Austin, *Report on the State of Public Instruction in Prussia*, 262.

165 Soon, students at these: Kaestle, *Joseph Lancaster*, 44–45.

165 "ended with a fizzle": Ibid., 44.

166 less efficiently than advertised: Esbjörn Larsson, "'Cheap, Efficient, and Easy to Implement'? Economic Aspects of Monitorial Education in Swedish Elementary Schools During the 1820s," *History of Education* 45, no. 1 (2016): 18–37.

166 nearly every state has: Common Core of Data, National Center for Educational Statistics; https://nces.ed.gov/ccd/schoolsearch/index.asp.

VII. OUTSIDE IN AND AT SCALE

171 The modern name was bestowed: Benjamin S. Bloom, "Learning for Mastery. Instruction and Curriculum. Regional Education Laboratory for the Carolinas and Virginia, Topical Papers and Reprints, Number 1," *Evaluation Comment* 1, no. 2 (1968).

171 Later, in 1919: Gloria Contreras, "Mastery Learning: The Relation of Different Criterion Levels and Aptitude to Achievement, Retention, and Attitude in a Seventh Grade Geography Unit" (Ed.D. diss., University of Georgia, 1975); https://files.eric.ed.gov/fulltext/ED111739.pdf.

171 mastery is experiencing: Kyle Spencer, "A New Kind of Classroom: No Grades, No Failing, No Hurry," *New York Times*, August 11, 2017.

174 Memory athletes: Martin Dresler, William R. Shirer, Boris N. Konrad, Nils C. J. Müller, Isabella C. Wagner, Guillén Fernández, Michael Czisch, and Michael D. Greicius, "Mnemonic Training Reshapes Brain Networks to Support Superior Memory," *Neuron* 93, no. 5 (2017): 1227–35.

174 In such fashion: Amir Raz, Mark G. Packard, Gerianne M. Alexander, Jason T. Buhle, Hongtu Zhu, Shan Yu, and Bradley S. Peterson, "A Slice of π: An Exploratory Neuroimaging Study of Digit Encoding and Retrieval in a Superior Memorist," *Neurocase* 15, no. 5 (2009): 361–72.

179 fewer than five hundred Americans: "Program Summary Report 2016," *AP Data—Archived Data 2016*, CollegeBoard.org; https://research.collegeboard.org/programs/ap/data/archived/ap-2016.

182 "building something that's easy": Coughlin, *The Longevity Economy*, 207.

182 In his 1910 book: Dewey, *How We Think*, 72.

182 Even traditional schools teach it: John L. Rudolph, "Epistemology for the Masses: The Origins of 'the Scientific Method' in American Schools," *History of Education Quarterly* 45, no. 3 (2005): 341–76.

184 "Zuckerberg-backed startup": Melia Russell, "Zuckerberg-Backed Startup That Tried to Rethink Education Calls It Quits," *San Francisco Chronicle*, June 28, 2019.

184 ammunition to critics: Audrey Watters, "The 100 Worst Ed-Tech Debacles of the Decade," *Hack Education*, December 31, 2019; http://hackeducation.com/2019/12/31/what-a-shitshow.

184 "It was disembodied and disconnected": Paul Emerich France, "Why I Left Silicon Valley, EdTech, and 'Personalized' Learning," *Reclaiming Personalized Learning*, January 15, 2018; https://paulemerich.com/2018/01/15/why-i-left-silicon-valley-edtech-and-personalized-learning/.

186 "Montessori saw in their grasping": Lillard, *Montessori*, 16.

187 "a general method": Piaget and Coltman, *Science of Education and the Psychology of the Child*, 147–48; Lillard, *Montessori*, 17.

187 The Montessori students outperformed: Angeline Lillard and Nicole Else-Quest, "Evaluating Montessori Education," *Science* 313, no. 5795 (2006): 1893–94.

187 in the 2017 study: Angeline S. Lillard, Megan J. Heise, Eve M. Richey, Xin Tong, Alyssa Hart, and Paige M. Bray, "Montessori Preschool

Elevates and Equalizes Child Outcomes: A Longitudinal Study," *Frontiers in Psychology* 8 (2017): 1783.

188 "high fidelity": Angeline S. Lillard, "Preschool Children's Development in Classic Montessori, Supplemented Montessori, and Conventional Programs," *Journal of School Psychology* 50, no. 3 (2012): 379–401.

189 *A Pattern Language:* Alexander, *A Pattern Language*, 424, 742.

191 As of 2019, Cambridge's public schools: *Cambridge Public Schools FY 2019 Adopted Budget*, prepared by the Cambridge, MA School Committee, Cambridge, MA, April 3, 2018.

VIII. TURN IT INSIDE OUT

198 The funny-sounding acronym: "A Brief History of MOOCs," McGill University, accessed November 11, 2019; https://www.mcgill.ca/maut/current-issues/moocs/history.

199 Battushig was one of them: Larry Hardesty, "Lessons Learned from MITx's Prototype Course," *MIT News*, July 16, 2012.

199 "I am, somehow, less interested": Stephen Jay Gould, "Wide Hats and Narrow Minds," *New Scientist*, March 8, 1979, 777.

200 As MOOCs' early cheerleaders: John D. Hansen and Justin Reich, "Democratizing Education? Examining Access and Usage Patterns in Massive Open Online Courses," *Science* 350, no. 6265 (2015): 1245–48; René F. Kizilcec, Andrew J. Saltarelli, Justin Reich, and Geoffrey L. Cohen, "Closing Global Achievement Gaps in MOOCs," *Science* 355, no. 6322 (2017): 251–52.

204 "the Internet of Things": Kevin Ashton, "That 'Internet of Things' Thing," *RFID Journal* 22, no. 7 (2009): 97–114.

204 I personally started up: Sanjay Sarma, "How Inexpensive RFID Is Revolutionizing the Supply Chain (Innovations Case Narrative: The Electronic Product Code)," *Innovations: Technology, Governance, Globalization* 7, no. 3 (2012): 35–52.

205 a nonprofit named Khan Academy: "What Is the History of Khan Academy?," Khan Academy; https://khanacademy.zendesk.com/hc/en-us/articles/202483180-What-is-the-history-of-Khan-Academy.

205 Stanford University: Andrew Ng and Jennifer Widom, "Origins of the Modern MOOC (xMOOC)," in *MOOCs: Expectations and Reality: Full Report*, by Fiona M. Hollands and Devayani Tirthali (Center for Benefit-Cost Studies of Education, Teachers College, Columbia University, 2014), 34–47.

206 From the start, it prioritized: David Kaiser, "The Search for Clean Cash," *Nature* 472, no. 7341 (2011): 30.

206 "The method of teaching": Richards, *Robert Hallowell Richards: His Mark*.

206 Instead, TEAL-style workshops: D. Fisher, Aikaterini Bagiati, and Sanjay Sarma, "Student Ambassadors: Developing an Older Student Cohort," in *Proceedings of the 40th SEFI Conference*, 2012.

209 Amol Bhave: Julie Barr, "Online Courses Paved the Way to MIT Graduation," *MIT News*, June 6, 2017.

209 "Nothing has more potential": Thomas L. Friedman, "Revolution Hits the Universities," *New York Times*, January 26, 2013.

210 by the end of 2018: Dhawal Shah, "By the Numbers, MOOCs in 2018," Class Central, December 11, 2018; https://www.classcentral.com/report/mooc-stats-2018.

210 Far more concerning: Hansen and Reich, "Democratizing Education," 2015.

210 MOOCs seemed to disproportionately benefit: Justin Reich and José A. Ruipérez-Valiente, "The MOOC Pivot," *Science* 363, no. 6423 (2019): 130–31.

210 In Agarwal's very first circuits course: Steve Kolowich, "The MOOC Survivors," *Inside Higher Ed*, September 12, 2012.

211 "dead": Varuni Khosla, "Udacity to Focus on Individual Student Projects," *Economic Times*, October 6, 2017; economictimes.indiatimes.com.

211 when one capital firm attempted: *Global EdTech Landscape 3.0*, Navitas Ventures, 2018; navitasventures.com/insights/landscape.

211 Clintondale High School: "Flipped Learning Model Dramatically Improves Course Pass Rate for At-Risk Students," Pearson Case Study, Pearson Education, 2013; http://assets.pearsonschool.com/asset_mgr/current/201317/Clintondale_casestudy.pdf.

211 Pamoja Education: Katrina Bushko, "Tackling Access to International Baccalaureate Courses with Blended Learning," Blended Learning Universe, May 15, 2018; https://www.blendedlearning.org/tackling-access-to-international-baccalaureate-courses-with-blended-learning.

211 "virtual schools": See, for instance, the Florida Virtual School: flvs.net.

212 a "flex" model: "Blended Learning Models," Blended Learning Universe; https://www.blendedlearning.org/models.

213 "deep learning" systems: Yann LeCun, Yoshua Bengio, and Geoffrey Hinton, "Deep Learning," *Nature* 521, no. 7553 (2015): 436–44.

214 a group of modern computer science luminaries: Iyad Rahwan et al., "Machine Behaviour," *Nature* 568, no. 7753 (2019): 477.

214 While the public was growing accustomed: Conner Forrest, "IBM Watson: What Are Companies Using It For?" ZDNet, September 1, 2015.

214 IBM announced Watson Education: "IBM Deal Expands Watson's Behind-the-Scenes Presence in Higher Education," edscoop, October 31, 2016; https://edscoop.com/ibm-deal-expands-watsons-behind-the-scenes-presence-in-higher-education.

215 The company Squirrel AI: Karen Hao, "China Has Started a Grand Experiment in AI Education. It Could Reshape How the World Learns," *Technology Review*, August 2, 2019.

216 An internal 2018 study: Chaitanya Ramineni and David Williamson, "Understanding Mean Score Differences Between the e-rater® Automated Scoring Engine and Humans for Demographically Based Groups in the GRE® General Test," *ETS Research Report Series* 2018, no. 1 (2018): 1–31.

219 In 1997, the situation was dire enough: Aallyah Wright and Kelsey Davis, "After Years of Inaction, Delta Teacher Shortage Reaches 'Crisis' Levels," *Mississippi Today*, February 18, 2019.

219 seven other approved online vendors: "Mississippi Online Course Approval (MOCA)," Mississippi Department of Education; https://www.mdek12.org/ESE/OCA.

219 "We're just going over": Aallyah Wright and Kelsey Davis, "Teacher Shortages Force Districts to Use Online Education Programs," *Mississippi Today*, February 20, 2019.

220 Roughly half of all people: Max Roser, Hannah Ritchie, and Esteban Ortiz-Ospina, "Internet," *Our World in Data*, 2019; ourworldindata.org/internet.

220 in India alone: Rishi Iyengar, "The Future of the Internet Is Indian," *CNN Business*, November 2018.

221 A sizable 2019 review study: Maya Escueta, Vincent Quan, Andre Joshua Nickow, and Philip Oreopoulos, *Education Technology: An Evidence-Based Review*, no. w23744 (National Bureau of Economic Research, 2017).

221 "isn't a story about an individual": Justin Reich, "The Village of the Boy Genius of Ulan Bator," *Education Week*, September 15, 2013.

223 it's hard to say exactly how prevalent: Zoe Kirsh, "The New Diploma Mills," *Slate*, May 13, 2017.

223 in 2011, 90 percent of U.S. school districts: Priscilla Rouse and Laurie Lewis, "Dropout Prevention Services and Programs in Public School

Districts: 2010–11. First Look. NCES 2011-037," *National Center for Education Statistics* (2011).

224 An important assumption: Lakoff and Johnson, *Metaphors We Live By*.

225 In a Circuits Analysis course: Candace Hazlett, "San Jose State University and edX Expand Course to CSU Campuses," edX Blog, November 13, 2013.

226 a concurrent experiment: Ry Rivard, "Udacity Project on 'Pause,'" *Inside Higher Ed*, July 18, 2013; Tamar Lewin, "After Setbacks, Online Courses Are Rethought," *New York Times*, December 10, 2013.

IX. THE SHOWDOWN

227 the Harvard economist Raj Chetty: Raj Chetty, Nathaniel Hendren, Frina Lin, Jeremy Majerovitz, and Benjamin Scuderi, "Childhood Environment and Gender Gaps in Adulthood," *American Economic Review* 106, no. 5 (2016): 282–88.

239 Not only did the blended students: Joshua Littenberg-Tobias and Justin Reich, "Evaluating Access, Quality, and Inverted Admissions in MOOC-Based Blended Degree Pathways: A Study of the MIT Supply Chain Management MicroMasters," SocArXiv preprint (2018).

EPILOGUE

260 "Every wheat farmer": Labaree, *Someone Has to Fail*, 54.

262 "Psychology is a science": James, *Talks to Teachers on Psychology: And to Students on Some of Life's Ideals*, 7–8.

Selected Bibliography

Aertsen, A., and V. Braitenberg. *Information Processing in the Cortex: Experiments and Theory*. Berlin, Heidelberg and Springer: 2012.

Alexander, Christopher. *A Pattern Language*. New York: Oxford University Press, 1977.

Altbach, Philip G. *Accelerated Universities: Ideas and Money Combine to Build Academic Excellence*. Leiden: Brill, 2018.

Bell, A. *An Experiment in Education, Made at the Male Asylum at Egmore, Near Madras: Suggesting a System by Which a School or Family May Teach Itself Under the Superintendence of the Master or Parent*. London: Cadell and Davies, 1805.

———. *The Madras School: Or, Elements of Tuition: Comprising the Analysis of an Experiment in Education, Made at the Male Asylum, Madras; with Its Facts, Proofs, and Illustrations; to Which Are Added, Extracts of Sermons Preached at Lambeth; a Sketch of a National Institution for Training Up the Children of the Poor; and a Specimen of the Mode of Religious Instruction at the Royal Military Asylum, Chelsea*. London: T. Bensley, 1808.

Benjamin, Aaron S., ed. *Successful Remembering and Successful Forgetting: A Festschrift in Honor of Robert A. Bjork*. New York: Psychology Press, 2011.

Bermúdez, José Luis. *Cognitive Science: An Introduction to the Science of the Mind*. Cambridge: Cambridge University Press, 2018.

Brown, Peter C., Henry L. Roediger, and Mark A. McDaniel. *Make It Stick: The Science of Successful Learning*. Cambridge, MA: Belknap Press, 2014.

Callahan, Raymond E. *Education and the Cult of Efficiency: A Study of the Social Forces That Have Shaped the Administration*. Chicago: University of Chicago Press, 1964.

Clandinin, D. Jean, and Jukka Husu. *The SAGE Handbook of Research on Teacher Education*. London: SAGE Publications, 2017.

Costandi, Moheb. *Neuroplasticity*. Cambridge, MA: MIT Press, 2016.

Coughlin, Joseph F. *The Longevity Economy: Inside the World's Fastest-Growing, Most Misunderstood Market*. New York: Public Affairs, 2017.

Cousin, V., and S. Austin. *Report on the State of Public Instruction in Prussia: Addressed to the Count de Montalivet*. London: E. Wilson, 1836.

Cuban, Larry. *Teachers and Machines: The Classroom Use of Technology Since 1920*. New York: Teachers College Press, 2004.

Dear, Brian. *The Friendly Orange Glow: The Story of the PLATO System and the Dawn of Cyberculture*. New York: Pantheon, 2017.

Dehaene, Stanislas. *Reading in the Brain: The New Science of How We Read*. New York: Penguin Books, 2010.

Dehn, M. J. *Working Memory and Academic Learning: Assessment and Intervention*. Hoboken, NJ: Wiley, 2011.

Dewey, John. "The Child and the Curriculum," 1902. https://archive.org/details/childandcurricul00deweuoft.

———. *How We Think*. [S.l.]: Project Gutenberg, 2011. http://www.gutenberg.org/files/37423/37423-h/37423-h.htm.

———. *The School and Society*. Chicago: University of Chicago Press, 1900.

Dewey, John, and J. A. Boydston. *The Early Works of John Dewey, 1882–1898, Volume 5: Early Essays, 1895–1898*. The Collected Works of John Dewey, 1882–1953. Carbondale: Southern Illinois University Press, 2008.

Ferster, Bill. *Teaching Machines*. Baltimore: Johns Hopkins University Press, 2014.

Fish, Jefferson M. *Race and Intelligence: Separating Science from Myth*. Mahwah, NJ: L. Erlbaum, 2002.

Gallistel, C. R., and Adam Philip King. *Memory and the Computational Brain: Why Cognitive Science Will Transform Neuroscience*. Chichester, UK: Wiley, 2011.

Gardner, Howard. *The Mind's New Science: A History of the Cognitive Revolution; with a New Epilogue by the Author; Cognitive Science After 1984*. New York: Basic Books, 1997.

———. *Multiple Intelligences: New Horizons*. New York: Basic Books, 2006.

Gatto, John Taylor. *The Underground History of American Education: A Schoolteacher's Intimate Investigation into the Problem of Modern Schooling*. New York: Oxford Village Press, 2003.

Gernsbacher, Morton Ann. *Psychology and the Real World: Essays Illustrating Fundamental Contributions to Society*. New York: Worth Publishers, 2011.

Gladman, F. J. *School Work*. Jarrolds' Teachers and Pupil Teachers Series, 1885.

Goldstein, Dana. *The Teacher Wars: A History of America's Most Embattled Profession*. New York: Anchor Books, 2015.

Gopnik, Alison, Patricia Kuhl, and Andrew Meltzoff. *The Scientist in the Crib: What Early Learning Tells Us About the Mind*. New York: Perennial, 2004.

Graebner, William. *A History of Retirement: The Meaning and Function of an American Institution, 1885–1978*. New Haven, CT: Yale University Press, 1980.

Gray, Peter. *Psychology*. New York: Worth Publishers, 2011.

Harrell, D. Fox. *Phantasmal Media: An Approach to Imagination, Computation, and Expression*. Cambridge, MA: MIT Press, 2013.

Huettel, Scott A., Allen W. Song, and Gregory McCarthy. *Functional Magnetic Resonance Imaging*. Sunderland, MA: Sinauer Associates, 2009.

James, William. *The Energies of Men*. New York: Moffat, Yard and Co., 1907.

———. *Psychology*. American Science Series, Briefer Course. H. Holt, 1893.

———. *Talks to Teachers on Psychology: And to Students on Some of Life's Ideals*. ATLA Monograph Preservation Program. H. Holt, 1899.

Jonçich, Geraldine. *The Sane Positivist: A Biography of Edward L. Thorndike*. Middletown, CT: Wesleyan University Press, 1968.

Kaestle, Carl F. *Joseph Lancaster and the Monitorial School Movement: A Documentary History*. Ed. with an Introduction and Notes by C. F. Kaestle. New York: Teachers College Press, 1973.

Kahneman, Daniel. *Thinking, Fast and Slow*. New York: Farrar, Straus and Giroux, 2015.

Kandel, Eric R. *In Search of Memory: The Emergence of a New Science of Mind*. New York: Norton, 2007.

Kandel, Eric R., and Sarah Mack. *Principles of Neural Science*. New York: McGraw-Hill Education, 2014.

Kazdin, Alan E. *Encyclopedia of Psychology. Aborti–System*. Vols. 1–7. Washington, DC: American Psychological Association, 2000.

Kimble, Gregory A., Michael Wertheimer, Charlotte L. White, and C. Alan Boneau. *Portraits of Pioneers in Psychology*. Vol. 1. Mahwah, NJ: Erlbaum/American Psychological Association, 1991.

Kliebard, Herbert M. *The Struggle for the American Curriculum: 1893–1958*. New York: RoutledgeFalmer, 2004.

Kushnir, Tamar, Janette B. Benson, and Fei Xu. *Advances in Child Development and Behavior*. Vol. 43. Amsterdam: Elsevier, 2012.

Labaree, David F. *Someone Has to Fail: The Zero-Sum Game of Public Schooling*. Cambridge, MA: Harvard University Press, 2012.

———. *The Trouble with Ed Schools*. New Haven, CT: Yale University Press, 2006.

Lagemann, Ellen Condliffe. *An Elusive Science: The Troubling History of Education Research*. Chicago: University of Chicago Press, 2009.

Lakoff, George, and Mark Johnson. *Metaphors We Live By*. Chicago: University of Chicago Press, 2017.

Lancaster, Joseph. *Epitome of Some of the Chief Events and Transactions in the Life of Joseph Lancaster, Containing an Account of the Rise and Progress of the Lancasterian System of Education . . . Written by Himself*. New Haven, CT: Baldwin & Peck, 1833.

Lillard, Angeline Stoll. *Montessori: The Science Behind the Genius*. New York: Oxford University Press, 2017.

Mahajan, Sanjoy. *Street-Fighting Mathematics: The Art of Educated Guessing and Opportunistic Problem Solving*. Cambridge, MA: MIT Press, 2010.

Meiklejohn, J. M. D. *An Old Educational Reformer, Dr Andrew Bell*. Edinburgh: W. Blackwood, 1881.

Metcalfe, Janet, and Arthur P. Shimamura. *Metacognition: Knowing About Knowing*. Cambridge, MA: MIT Press, 1995.

Murchison, Carl, ed. *A History of Psychology in Autobiography*. Vol. 3. Worcester, MA: Clark University Press, 1936.

Newmeyer, Frederick J. *The Politics of Linguistics*. Chicago: University of Chicago Press, 1986.

Papert, Seymour. *The Children's Machine: Rethinking School in the Age of the Computer*. New York: Basic Books, 1994.

———. *Mindstorms: Children, Computers, and Powerful Ideas*. New York: Basic Books, 1993.

Piaget, Jean. *Science of Education and the Psychology of the Child*. Trans. Derek Coltman. New York: Viking, 1970.

Ravitch, Diane. *Left Back: A Century of Battles over School Reform*. New York: Touchstone, 2001.

Renninger, K. Ann, and Suzanne Hidi. *The Cambridge Handbook of Motivation and Learning*. Cambridge: Cambridge University Press, 2019.

Richards, Robert H. *Robert Hallowell Richards: His Mark*. Boston: Little, Brown, 1936.

Shenk, David. *The Genius in All of Us: New Insights into Genetics, Talent, and IQ*. New York: Anchor Books, 2011.

Skinner, B. F. *About Behaviorism*. London: Penguin Books, 1974.

Southey, R. *The Origin, Nature, and Object, of the New System of Education*. London: J. Murray, 1812.

Southey, R., C. C. Southey, and C. B. Southey. *The Life of the Rev. Andrew Bell: . . . Prebendary of Westminster, and Master of Sherburn Hospital, Durham. Comprising the History of the Rise and Progress of the System of Mutual Tuition . . .* London: J. Murray, 1844.

Thorndike, Edward L. *Educational Psychology*. New York: Teachers College, Columbia University, 1913.

———. *Human Nature and the Social Order*. New York: Macmillan, 1942.
———. *Notes on Child Study*. Columbia University Contributions to Philosophy, Psychology and Education. New York: Macmillan, 1901.
———. *The Principles of Teaching: Based on Psychology*. New York: Taylor & Francis, 2013.
———. *The Psychology of Learning*. Educational Psychology. Teachers College, Columbia University, 1913.
Tobias, Sigmund, and Thomas M. Duffy. *Constructivist Instruction: Success or Failure?* London: Routledge, 2009.
Toffler, Alvin. *Future Shock*. London: Pan Books, 1994.
Tschurenev, Jana. *Empire, Civil Society, and the Beginnings of Colonial Education in India*. Cambridge and New York: Cambridge University Press, 2019.
Turbayne, Colin Murray. *The Myth of Metaphor*. New Haven, CT: Yale University Press, 1962.
Tyack, David B. *The One Best System: A History of American Urban Education*. Cambridge, MA: Harvard University Press, 1976.
Tyack, David, and Larry Cuban. *Tinkering Toward Utopia: A Century of Public School Reform*. Cambridge, MA: Harvard University Press, 1995.
Varela, Francisco J., Eleanor Rosch, and Evan Thompson. *The Embodied Mind: Cognitive Science and Human Experience*. Cambridge, MA: MIT Press, 2017.
Vygotsky, Lev Semenovich, and Michael Cole. *Mind in Society: The Development of Higher Psychological Processes*. Cambridge, MA: Harvard University Press, 1979.
Westbrook, R. B. *John Dewey and American Democracy*. Ithaca, NY: Cornell University Press, 2015.
Wilson, Donald A., and Richard J. Stevenson. *Learning to Smell: Olfactory Perception from Neurobiology to Behavior*. Baltimore: Johns Hopkins University Press, 2006.
Zimmerman, Barry J., and Dale H. Schunk. *Educational Psychology: A Century of Contributions; A Project of Division 15 of the American Psychological Society*. New York: Routledge, 2014.

Index

About the Authors

Sanjay Sarma is Vice President for Open Learning at MIT. A professor of mechanical engineering, he has worked in the fields of education, computational geometry, manufacturing, radio frequency ID, sensors, and automotive technology. He has advanced degrees from Carnegie Mellon and UC Berkeley.

Luke Yoquinto is a science writer and a researcher at the MIT AgeLab. His work can be found in such publications as *The Washington Post*, *Slate*, *The Wall Street Journal*, and *The Atlantic*. He is a graduate of Boston University's science journalism program.